D1930120

The Way of the Carpenter

THE WAY OF THE CARPENTER

TOOLS AND JAPANESE ARCHITECTURE

by William H. Coaldrake

New York WEATHERHILL *Tokyo*

First edition, 1990

Published by Weatherhill, Inc., New York, with editorial offices at Tanko-Weatherhill, Inc., 8–3 Nibancho, Chiyoda-ku, Tokyo 102, Japan. *Copyright © 1990 by William Howard Coaldrake; all rights reserved.* *Printed in Japan.*

Library of Congress Cataloging in Publication Data: *Coaldrake, William Howard. / The way of the carpenter: tools and Japanese architecture / p. cm. / Includes bibliographical references. / 1. Carpentry—Japan—Tools—History. I. Title. / TH5618.C63 1990 / 694.0952—dc20 / 90-35137 CIP / ISBN 0-8348-0231-7*

Contents

Preface

As a child of missionary parents living in Izu, Japan, in the Occupation years, I can scarcely claim to have been born with a silver spoon in my mouth. However, I can claim to have had Japanese carpenters' tools in my hands from early childhood. In those days of desperate and rapid rebuilding of a whole nation, camps, churches and houses were springing up throughout the Izu Peninsula under the direction of my father, who was both carpenter and priest. My earliest memories, therefore, are of building sites. I have a clear picture as a three-year-old of clambering around the roof frame of the new rectory in Itō. The fragrance of unseasoned timber was as familiar to me as the odor of the "honey buckets" of the night-soil carriers in the neighboring fields.

It may have been this childhood experience which drew me inexorably over the years of scholarly study and research in art and architecture to the building sites of Japan and the patient but energetic world of the master builders and restorationists. Their enterprises may be eclipsed by the onslaught of high technology, "intelligent" buildings and manufactured housing in the vast urban developments of Japan today, but their practice continues in the ancient architectural centers of Kyoto and Nara.

In 1979, when still a graduate student in fine arts at Harvard University, I became a part of this special world by working with traditional Japanese carpenters in the reconstruction of an eighteenth-century Kyoto merchant's shop and residence at the Boston Children's Museum. The same master carpenters took me under their tutelage when I returned to Japan in 1980–81 to work at Enryaku-ji, the great Buddhist temple complex among the cedars of Mount Hiei, overlooking Kyoto. Much of the Kakushō-an, a subprecinct, was being rebuilt by carefully coordinated teams of master carpenters (*daiku*), roof tilers, wall plasterers and stone masons. Under their guidance, I became aware of the extraordinary complexity of construction processes and the infinitely varied techniques for using traditional tools. I was the first non-Japanese admitted into the membership of the Kyōto Dentō Kenchiku Gijutsu Kyōkai (Kyoto Guild of Traditional Master Builders). Subsequently, I continued my involvement with the hereditary master builders, and the study of their tools and techniques, which has made this book possible.

In this book I have combined a discussion of tool evolution with a technical analysis of tool types and techniques in the hope of narrowing the gap between these two fields which may sit awkwardly in each other's company. I have tried to be both artisan and academic. A technical study gives little sense of why tools were invented and how they affected building form. A purely academic approach overlooks the logic of practical problems and their resolution. Only a dual approach meets both challenges because the tools themselves are the link between practice and principle in the Way of the Carpenter.

I am deeply indebted to Yasui Kiyoshi, master carpenter and restorer of Katsura Imperial Villa, for his assistance; to Inagaki Eizō, professor emeritus of the Department of Architecture of the University of Tokyo; Karai Kunio, formerly of the Takenaka Carpentry Tools Museum in Kobe; and to John M. Rosenfield, Abby Aldrich Rockefeller Professor of Oriental Art at Harvard University. I have learned much from the many artisans and craftsmen alongside whom I have worked, as well as from the tool collection and curators of the Takenaka Carpentry Tools Museum, the major repository of carpenters' tools in Japan. I am also grateful to The Japan Foundation for a Professional Fellowship in 1986–87.

Conventional Japanese proper name order, family name first, is followed. Macrons have been omitted from common nouns like Shinto and daimyo which appear in modern English-language dictionaries and which are not italicized. Special terms are italicized on first appearance; thereafter they are not. Traditional measurements retain italics for clarity. All translations from Japanese as well as the line drawings and photographs are my own unless otherwise acknowledged. A glossary giving the full name and Japanese characters for technical terms used in this study is also appended.

PHOTO CREDITS

Coaldrake, Frank W.: figs. 95, 100
Hosoda Inc.: fig. 14
Imperial Household Agency: pl. 2; figs. 15, 126
Kimura Ethnology Museum: fig. 133
Kintetsu Historical Museum: fig. 31
Kita-in: fig. 129
Kusadosengenchō Iseki Chōsa Kenkyūjo: fig. 127
Kōsaku-sha: fig. 23
Mie-ken Maizō Bunkazai Center: fig. 128
Miyoshino Shrine: pl. 6
Shinshō-ji: fig. 4
Sōshi-sha: figs. 112, 117
Suntory Museum of Art: pl. 5
Suō Tenman-gū: pls. 3, 4; fig. 6
Takenaka Carpentry Tools Museum: figs. 3, 5, 22, 24, 25, 27, 29, 30, 32, 34, 35, 37, 38, 42, 49, 51, 55, 56, 61, 62, 64, 65, 70, 71, 78, 79, 81, 83, 84, 85, 86, 87, 89, 91, 92, 94, 96, 99, 137, 138, 139, 141, 143
Tokyo National Museum: fig. 116

Pl. 1. *Sumitsubo* (ink pot and snap line). Ca. 1800. Carved by a master carpenter from a single block of *keyaki* (zelkova). The small figure is a carpenter using the snap line to draw a line on a board in preparation for cutting. *Takenaka Carpentry Tools Museum.*

Pl. 2. *Kasuga gongen genki-e.* Scroll 1. 1309. Detail of shrine building site. *Imperial Household Agency.*

Pl. 3. *Matsuzaki tenjin engi emaki*. Scroll 4. 1311. Detail of shrine construction. *Suō Tenman-gū*.

Pl. 4. *Matsuzaki tenjin engi emaki*. Scroll 6. 1311. Detail of shrine construction. *Suō Tenman-gū*.

Pl. 5. Two-man frame-saw (*oga*) used for ripsawing lumber as shown in *Sanjūniban shokunin uta-awase*, ca. 1500. *Suntory Museum of Art.*

Pl. 6. *Kawagoe miyoshino tenjin engi emaki*. Mid-seventeenth century. Detail showing seated use of block planes. *Miyoshino Shrine.*

Pl. 7. Adjustable chamfer planes (*jiyū-mentori-ganna*).
Takenaka Carpentry Tools Museum.

Pl. 8. Plough planes (*sokotori-jakuri-ganna*) for cutting grooves for sliding screens. Both planes have been "remouthed" with a hardwood block after extensive use. *Takenaka Carpentry Tools Museum.*

Pl. 9. Wide-bladed butt chisel (*hiro-nomi*) made of watered steel. Blade head 4.85 cm. wide (1 *sun* 6 *bu*). Made ca. 1930 by Zensaku II. *Takenaka Carpentry Tools Museum.*

Pl. 10. Seventeenth-century one-man wide-bladed ripsaw (*maebiki-oga*). *Takenaka Carpentry Tools Museum.*

The Way of the Carpenter

MAJOR PERIODS IN JAPANESE HISTORY

Jōmon	ca. 10,000 B.C.–ca. 300 B.C.
Yayoi	ca. 300 B.C.–ca. A.D. 300
Tumulus	ca. 300–ca. 700
Asuka	552–710
Nara	710–794
Heian	794–1185
Kamakura	1185–1333
Muromachi	1333–1573 (Civil War period: 1467–1573)
Momoyama	1573–1603
Edo (Tokugawa)	1603–1867
Meiji	1868–1912
Modern Era	1912–present
Taishō	1912–26
Shōwa	1926–89
Heisei	1989–Present

JAPANESE TRADITIONAL UNITS OF MEASUREMENT

Linear (as standardized in 1886)

1 *rin*	=			0.303 mm (0.012 inch)
10 *rin*	=	1 *bu*	=	3.03 mm (0.12 inch)
10 *bu*	=	1 *sun*	=	3.03 cm (1.2 inches)
10 *sun*	=	1 *shaku*	=	30.3 cm (0.994 foot)
6 *shaku*	=	1 *ken*	=	1.82 m (1.99 yards; originally one span)
10 *shaku*	=	1 *jō*	=	3.03 m (3.31 yards)

Square

1 *tsubo*	=	3.305 m^2 (3.952 square yards; originally the area of two tatami mats laid side by side)

CHAPTER ONE

The Way of the Carpenter

Three small, unassuming Japanese planes rest on my desk. The gift of a retired master carpenter, none is larger than the span of my outstretched fingers nor greater in width than the base of my thumb. Two are straight, one sweetly curved. The oak block of each plane is scarred with use, the cutting blade pockmarked with rust. In such objects John Ruskin may have glimpsed the romance of ruin but these tools evoke more than nostalgia for a disappearing craft. They form tangible links with the people and processes of building in the past. They are not passive objects meant for casual perusal in a museum. They were the instruments of energetic mediation between man and materials in a world which has now all but disappeared. Pulled across a piece of wood, each of my planes still carves its distinct signature into the grain—a neat groove, a serrated path or a graceful arc. The true home of such tools is not a desk top nor a display case but in the hands of a carpenter standing among the fresh shavings of a building site. They belong to the dynamic world of traditional Japanese building practice as an inseparable part of a way of work which is also a way of life. This is the Way of the Carpenter.

The simple word "way" has a wide range of meanings. A way is the method or manner of doing something. On a grander scale it may be a custom or convention upon which a society operates. A way serves as a direction, route, course or path. It forms patterns of thinking and action. Jesus of Nazareth pronounced himself "the Way, the truth and the life" while Lao-tse made the Tao, or the Way, the basis of social and governmental ethics. In Japan, the concept of way (*michi* or *dō*) is manifest in many aspects of culture and belief. Shinto is the Way of the Gods and *Butsudō* the Way of the Buddha. Bushido, the Way of the Warrior, has passed into the universal vocabulary of the martial arts; these include aikido, judo and kendo. Tea ceremony is known as *chadō*, the Way of Tea, calligraphy as *shodō*, the Way of Writing, and flower arrangement is *kadō*, the Way of the Flower.

The concept of Way is crucial for understanding carpenters and their tools in Japan. The term *dōgu*, which is most commonly used for tools in Japan, means literally "the way of the tool." It reflects the association of the process of building with a broader pattern of meaning best defined by the concept of Way. Dōgu consists of two characters 道具. *Dō* means "way," and *gu* is best translated as "implement" or "means." When a carpenter's chisel or plane is called a dōgu, therefore, it is being recognized as an implement or a means of a special Way. When the Chinese characters for dōgu were adopted into the Japanese language many centuries ago, they did not refer to carpenters' tools at all. They designated the important implements used in Buddhist ritual as aids to enlightenment. Moreover, they referred to the clothing and other equipment essential to the priestly way of life, such as the prayer mat used for meditation and the bowl used for begging. When the samurai emerged as a ruling class in the twelfth and thirteenth centuries, the meaning of the term broadened to include many of the special items that were part of their way of life and warrior creed including armor, spears and sword sheaths. By the fourteenth and fifteenth centuries, the meaning of dōgu had widened to include the tools of the carpenter.

The adoption of a word with deep religious associations like dōgu to designate carpenters' tools is neither sacrilege nor a violation of sacred tenets. Historically, Japanese carpenters were far more than manual workers because the process of designing and erecting a building was more than a routine construction procedure. There is a universal association between the process of creating buildings and the spiritual dimension of the human psyche. In Ruskin's words, "great architecture makes us believe what we would not otherwise have believed."[1] The creativity necessary for sustaining the process of construction is sparked by the same impulses of imagination and awe as fire spiritual beliefs, and the process of building shares many of the same attributes as religion, including sequence, order, orthodoxy, hierarchy, organization and awareness of higher imperative. The construction of buildings dedicated to higher purposes of religion on one level, and to the creation of metaphors of faith on another, may guide the initiate along the path of inner development as well as sharpen the definition of a faith. It is no accident that religion makes common use of architectural imagery such as the House of God and the Gateway of Heaven, and that the symbols of Freemasonry are tools of the medieval builder.

The process of building in Japan thus entailed far more than the procedures of erecting and decorating structures. It became a special way of life, a pattern of existential belief which, over the course of many lifetimes, took on the richness of the term Way. This is seen in the observance of complex rituals based on Shinto rites throughout the process of construction, in the esoteric nature of the organization and training of the profession of the carpenter, and in his rituals and reverance for tools. Let us see how each of these aspects contributes to the Way of the Carpenter.

SHINTO RITES AND THE WAY OF THE CARPENTER

The deep religious meaning of the Way of the Carpenter is reflected in the many

FISCHBEK :: Gleich drei Titel erkämpften die Boxer des TV Fischbek bei den Hamburger Meisterschaften der Jugend vor 400 Zuschauern in der Boxsporthalle Hamburg am Braamkamp. Farhad Noorrie gewann im Finale der Gewichtsklasse bis 60 Kilogramm gegen Farsad Rezae (Bonsai Hamburg), Yunus Öztürk schlug in der Klasse bis 52 Kilogramm Grisho Shmavonyan (Agon Hamburg) und Vladislav Kiselev (bis 44 Kilogramm) siegte gegen Tyron Duncan-Williams (Agon). Alle drei ge-

SV mit 0:1 und verabschieden sich

; that have been performed since ancient times as ; process. Most of these ceremonies follow the rites ı, but there are wide variations in observances customs. Examples of Shinto rites to be observed e Grand Shrines of Ise are contained in the *Engi-* 'n records of imperial court etiquette, compiled in *gi-shiki* includes detailed lists of the offerings to be *kami* on such important occasions as the Pillar- npletion Ceremonies. The cyclic rebuilding of the enty years is the supreme example of architecture y-two major ceremonies and festivals are spread actual building operation, ranging from the 's offered to the kami of the mountain where the he construction will be felled, to the solemn e sacred objects from the old shrine buildings to ıries. Every stroke of an adze and every cut of a saw is presided over or executed by carpenters who have been specially purified and who wear white clothing to avoid defilement of the sacred site.

Even in contemporary Japan, ground-breaking, ridge-raising and completion ceremonies are still observed in many building projects. The ground-breaking ceremony (*jichinsai*) begins the on-site construction process with rituals designed to appease the tutelary gods of the building site and to pray for the safety of all concerned in the building operation. Young bamboo stalks are used to mark the four corners of a sacred enclosure on the building site. These are joined by a sacred rope of new straw (*shimenawa*) bearing sacred, folded white invocatory papers. An altar is erected with a separate table for offerings of rice, salt, fruit and sakè. Blessings and purification rites are performed with a switch of leaves of *sakaki* (*Cleyera japonica*), the sacred Shinto tree.

The most important ceremony, celebrating completion of the assembly of the timber-frame, is the ridge-raising ceremony (*jōtōshiki* or *muneageshiki*). It includes the ceremonial raising and positioning of the ridgepole, thanksgiving for the safe erection of the frame, and prayers for the long life and well-being of the building and its inhabitants. The official ceremonies are regularly followed by feasting and celebration. The completion ceremonies (*rakuseishiki*), which involve all participants as well as neighbors and friends in prayer, feasting and toasting, are less rigid in form.

In traditional practice, the chief master carpenter himself frequently officiated at the ceremonies, donning the robes of a Shinto priest and intoning invocations and blessings ancient in origin (figs. 1 and 2). The chief master carpenter was formerly responsible for making the ritual implements and decorations for these ceremonies, such as the ridge plaque and sacred bow and arrow that were attached to the ridgepole during the ridge-raising ceremony, but this practice has largely lapsed (fig. 3).

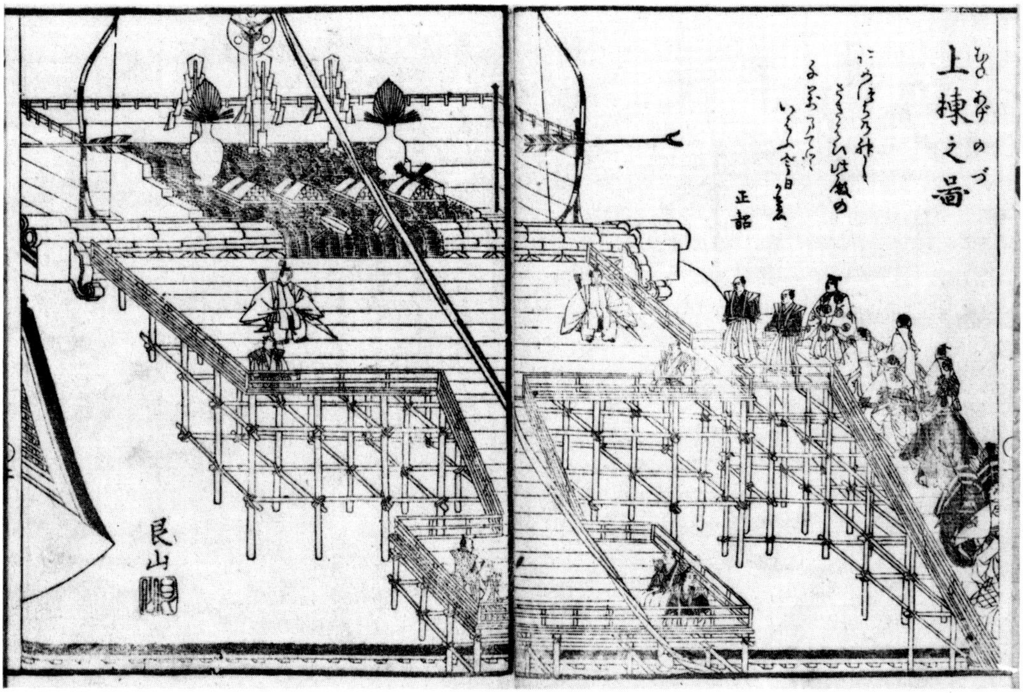

Fig. 1. Carpenter's manual showing the correct way of performing the ridge-raising ceremony (*Shōke hitsuyōki*, 1756).

Fig. 2. Ridge-raising ceremony in the restoration rebuilding of Katsura Imperial Villa, 1979.

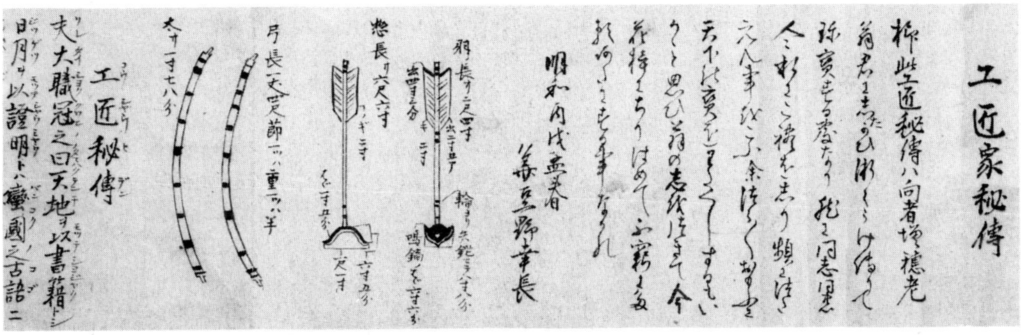

Fig. 3. Secret scrolls giving instructions for the way of making and using the sacred bows and arrows at a ridge-raising ceremony (*Kōshōke hiden*, 1772).

THE MAN AND THE WAY—THE MAKING OF THE CARPENTER

The daily life and activities of the carpenter, and the organization of the building profession generally, had much in common with that of the craft guilds of medieval Europe. They entailed careful management of relationships with patrons and clients, healthy competition with other workshops and intensive training of the next generation of carpenters. The Japanese carpentry profession, like the guilds of Europe, reflected religious piety and ritual in its professional activities, having much of the character of an enclosed religious order. The kami who watched over the carpenters were "kindred spirits" to the patron saints of the European building guilds. A meticulous process of selection and initiation into its various secret practices lay at the basis of the life and strength of the carpentry profession. The accumulated wisdom of generations of experience in architectural style and construction procedures was enshrined in secret hereditary writings. These ensured conformity to the articles of faith in their own design tradition from one generation to the next within each workshop.

Carpenters followed the customary pattern of training in which manual skills were mastered within a disciplined way of life governed by a strict sense of ritual and hierarchy. Reverence for tradition was paramount. The body of precepts and practices of one's predecessors in the tradition were held in sacred trust to be mastered, followed and passed on intact to the next generation. Carpentry was as worthy of devotion as any religious faith and never lacked suitable aspirants to its life of discipline and devotion.

Boys in their thirteenth year were considered to be sufficiently endowed both physically and spiritually to undertake the contractual terms of service of apprentices, furnish a guarantor and abide by the required solemn vows to master the techniques and maintain the secret skills of the particular workshop or "house" to which they were admitted. The master, for his part, guaranteed the provision of proper food and clothing and certain other small bounties such as New Year gifts. However, his side of the agreement was less strictly regulated than that of the apprentice, and though the guilds officially adopted a stance of tacit oversight of the appointment and training of apprentices, it was left largely to the good sense and good will of the individual guild member to care for his apprentices in an appropriate manner. In the regulations of one guild of carpenters in the city of Edo (Tokyo) for 1793 there is the stipulation concerning apprentices that "even in cases of minor misdemeanors they are not to be treated unjustly and they are to be given a sound training," which is more a general exhortation of principle than detailed regulation.[2]

In view of the totality of their commitment, it is surprising to find that there are no lists of duties, gradations of study and learning, or regulations in manual instructions as might be found in any school curriculum, especially as the shogunate had embarked on systematic propagation of Neo-Confucian tenets of education in the late seventeenth century. It is important, however, to bear in mind that a high level of literacy was assumed in the case of the carpenters who rose to the top of their profession, and this was obtained within the apprenticeship system. In some workshops, as in the commercial houses in the great cities of Edo and Osaka in the seventeenth century, reading and calligraphy classes were conducted for apprentices by masters after the day's duties were finished. Toward the end of Tokugawa rule, the widespread accessibility of *terakoya*, the area or village Buddhist temple school, to all children desiring instruction in ethics, calligraphy and abacus meant that a lad of twelve or thirteen years often had some such skills at the time of his acceptance as an apprentice.

In a typical ten-year apprenticeship, the first few years would be fully occupied with miscellaneous chores in the house workshop. A young apprentice was forbidden to touch any tools, let alone receive instruction in their use during that time. This applied equally to the older and often more educated apprentices who had started their training late. The apprentice was expected to watch the master and his assistants at work until such time as he was overcome with an overwhelming desire to use the tools himself. Even then, only the most basic techniques were taught directly. Skills of a more advanced nature were absorbed by imitation of the master by the apprentice, a process known universally throughout the arts and crafts of Japan as *nusumi-geiko*, "stolen lessons." In traditional practice, therefore, the bases of learning were total immersion in a workshop environment and inner motivation, both occurring at an age of extreme sensitivity and adaptability. The manual skills and creative outlook of the apprentice were formed even as the very person was maturing. The essence of carpenter education was to mold the person and the professional at the same time, before "the bones hardened." Having been exposed to the subtle psychology of this system, I know that its techniques establish a progression defined not by formal curriculum but by behavioral initiation, and that the whole being is permeated with special attitudes toward collaborative process while the techniques of tool use and building construction are being mastered.

After completing ten years as an apprentice, the aspiring carpenter could be presented by his master to other craftsmen in a formal ceremony in which he received a set of clothes and a set of tools. In theory he was now ready to receive full wages as a craftsman or to set up in the profession for himself, but in practice it was customary to serve a further period with his master as a thanks offering. Even after independence, the familial nature of the master-apprenticeship relationship continued for a lifetime with annual exchange of visits and gifts.

THE RITUALS OF TOOLS

In the most immediate and practical sense the Way of the Carpenter was also the Way of the Tool. It was the method and manner of selecting timbers, determining their dimensions and fashioning and assembling them. The carpenter created and used a variety of tools in the accomplishment of these tasks. There were tools for designing the building, which may be broadly categorized as "instruments of the mind," and there were tools for cutting and finishing the timbers, or the "tools of the hand." This second category also included special tools used for maintaining the tools themselves.

The procedures of tool use were the fundamental means for realizing architectural form. The interaction between the carpenter and his tools and, in turn, between his tools and the materials, formed the basis for the customs and conventions evolved in countless family workshops. Many of the ceremonies and rituals observed as part of the building process were specifically devoted to tools. Although rarely seen today, special rites were performed to bless and purify the inking of the first marking line, the first stroke of the adze and the first stroke of the saw. According to the *Engi-shiki*, by the tenth century the consecration offerings at the periodic rebuildings of the Ise Grand Shrine included planing-knives, adzes, chisels and saws along with hen's eggs, chickens, sickles and mattocks. Simpler versions of these same rituals are still observed in remoter rural areas of Japan today, where it is customary for farmers to affix building tools to votive paintings at local shrines as an offering to the gods, along with agricultural implements like sickles and other appropriate rustic offerings. Some votive paintings even depict carpenters' tools like the ink pot and snap line (fig. 4), a logical extension of their adoration at local festivals and religious rites. The statutory blessing of tools as part of religious rituals performed over many generations at building sites, together with practices like the offering of tools to local gods, bestowed on them a sacred meaning. It also led to the creation of ceremonial tools reserved for ritual purposes. Each family of carpenters had a special set of adzes, saws, ink pots and squares which were reserved for use at the time of the ridge-raising ceremony. These tools, with their cutting edges dull but their handles bright with gold and lacquer, were purified and blessed as part of the rites dedicating the new building (fig. 5).

Tools with special religious and symbolic associations have thus been central to the profession of the building carpenter. However, the tools of daily use were also imbued with special significance, as we have seen, and all the organizational, vocational and religious dimensions to the Way of the Carpenter were focused on the actual process of making and using these tools. The handles of chisels and saws, and the wooden blocks of planes, were made by the carpenter himself to suit the feel of his own hand and the balance of his body

Fig. 4. Votive painting (*ema*) of carpentry tools. 1814. *Shinshō-ji, Narita.*

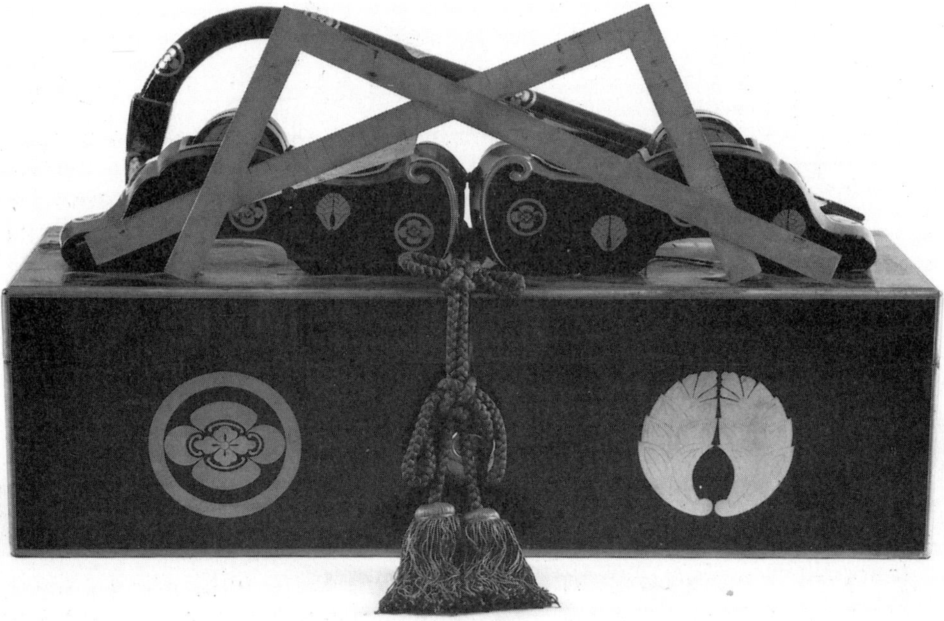

Fig. 5. Eighteenth-century set of ceremonial carpenter's tools.

and mind. The steel cutting blades were fashioned by a smith working to the carpenter's specifications. Tools so formed from a blend of personality and material technology became the embodiment of the carpenter's technical skills and an expression of his personality. He cherished and cared for them as if they were his own children. He addressed them with the respect accorded his parents. He spent many hours each day patiently and expertly sharpening their cutting edges. The greatest care was always taken in their handling and storage. One of the last traditionally trained surviving master temple carpenters, Nishioka Tsunekazu, recalls:

> When my uncle was still young he would put each tool carefully into its own special oiled-paper pouch before returning it to the box . . . . He would treat his planes like precious hanging scrolls, placing each one in its own separate pouch. He put chisels in their own special sheaths before returning them to the toolbox. These days only the most important paring chisels are put in their own sheaths—the rest are just dropped back in as they are . . . .[3]

The carpenter could follow the right path only if he used the appropriate tool in the correct way for each task. With these tools he came to challenge the very limits of materials and techniques. Some of the most elaborate, self-conscious joinery found in premodern buildings is no mere subservience to structural necessity. It is joyful exaltation in perfection of craft as art and the perfection of self.

THE SECRET WORLD OF THE MASTER CARPENTER

Ritualistic practices explain some of the challenges and problems encountered in the study of Japanese carpenters' tools. It is difficult for the outsider to gain entry into the secret world of the profession that uses and cherishes them. Historically, the veil of secrecy descended when it came to the transmission of the specific carpentry techniques within a particular workshop or house. These techniques belonged to a secret and circumscribed world to which admission was frequently denied even to other members of the same guild. There are a number of instances where guild regulations specify that "no apprentices belonging to other domains (*han*) are to be accepted" to ensure that the secrecy of teachings was not violated. The esoteric nature of workshop practices was thus reinforced by craft guild sanctions. The same type of secrecy prevailed in the European guilds. Even today a strictly trained hereditary carpenter never shows his tools to the uninitiated because these tools are the embodiment of the techniques and the traditions entrusted to him by his forebears. There are carpenters who will display with cheerful readiness an impressive collection of shiny and attractive tools. The real work, however, is done with other tools never exposed, brought out only when they are called upon to perform their appointed task—like the samurai sword which is never drawn except for the killing stroke of the legendary one-stroke swordsman. Each tool has its strictly defined moment and place of use. It is treated with the easy familiarity of long acquaintance when put to proper use and with something bordering on religious devotion when put to rest.

Improper use of tools or insensitive questions and photography constitute a desecration of a sacred trust and tradition. To the extent that I have been initiated into the world of architecture and the Way of the Carpenter as a member of the modern guild for traditional building craftsmen, the Kyōto Dentō Kenchiku Gijutsu Kyōkai, I am obliged to maintain the traditions and respect its esoteric rituals as both profession and faith.

Carpentry tools, therefore, lie at the heart of the vocational practice and spiritual beliefs of a complex profession. Tools assume their full and proper stature, their rightful and complete meaning, only when understood as part of the profession they serve as instruments of the Way of the Carpenter.

CHAPTER TWO

Men, Materials and Methods

> If it wishes to avoid the boredom of fixed abstractions, the book of tools will also be the book of materials and trades.
>
> —André Velter and Marie-José Lamothe, *Le livre de l'outil* [1]

There are four keys to understanding the Way of the Carpenter. The first is the carpenter himself, the particular skills and the orderly structure of his profession. The second is the material he works, the inherent properties and attributes of the different woods which form the structure and much of the detailing of the Japanese building. The third is found in his methods of building design and construction. The fourth key is his set of tools, the dōgu, which link man, materials and methods.

It is essential to examine the first three aspects before explaining the fourth, the carpenter's tools themselves.

THE PROFESSION OF THE CARPENTER

Who, then, is this person, the Japanese carpenter? Is he an artisan or an artist, architect or builder? What does he call himself? How does he organize his profession?

The essential point to make is that the carpenter is only one of the many craftsmen engaged in the building process. Japanese architecture is the composite of multiple trade traditions, each of which developed over many generations. In addition to the carpenters there were stone masons, plasterers, roof tilers, copper sheet roofers, thatchers and shinglers. There were also the *tateguya*, makers of shoji sliding screens and sliding door frames for the thick paper screens, as well as the makers of tatami mats (*tatamiya*), each an indispensable part of the residential interior. The cabinet makers, an allied profession, crafted the chests (*tansu*) and other interior furnishings while the box maker (*sashimonoshi*) fashioned containers of paulownia wood for storing works of art and daily utensils. Fellers and sawyers were responsible for the harvesting and transportation of trees as well as their metamorphosis into workable timbers.

These trades constitute the inner circle of the building professions. The pattern of professions involved in the building process spread like the ripples on the surface of a pond. Master smiths specialized in forging the laminated steel blades for planes, chisels and saws. They also manufactured mountains of iron nails and clamps needed to strengthen the timber-frame structure. Makers of mulberry paper provided the coverings for the shoji screens, the width of the paper helping to determine the spacings of the frame made by the screen makers. Ornamental metal workers, painters, sculptors and lacquer craftsmen each added their own special dimension to the finish of the building.

These professions were organized in similar hereditary and behavioral patterns, but each was autonomous in administration and activity. Carpenters and roof tilers still sit in separate groups around the communal fire for morning teatime at wintry building sites, long accustomed to each other's presence at the building site but exchanging little more than formal courtesies with each other.

In the pyramid of building professions and allied craft trades, it was the carpenter who was preeminent. This was not due to any obvious difference in status or pay scale but simply because the upper echelon of master carpenters was responsible for the overall design and execution of the building. The work of the other professions in all respects conformed to the master plan devised by the chief master carpenter. The carpenter has always projected a powerful presence at every building site. He has left the indelible stamp of his personality on each building he has constructed and each tool he has used. We may not always know his name but we can sometimes recognize the special techniques of the workshop to which he belonged. Numerous written and pictorial sources also help us to identify him. Moreover, most buildings of consequence had a wooden plaque (*munafuda*) attached to the ridgepole listing information about the builders, particularly the carpenters, who participated in the project. Government documents from the Edo period record with scrupulous detail the organization and status of the vast armies of carpenters working in official service. Families such as the Heinouchi and the Kōra rose from artisan status to become, in effect, official court architects, creating some of the most lavish and impressive monuments ever built.

The carpenter, therefore, must be seen as both artisan and artist, architect and builder. Like the medieval mason of Europe, the Japanese carpenter designed and built in a single, organic process which will be described later in the chapter. It is important to note here that this process is very different from architectural practice today. The modern architectural profession makes a doctrinaire distinction between design and execution. This is reinforced by separate training for architects and builders, making it difficult to appreciate the role of the traditional master carpenter who did both.

The Japanese word *daiku*, the name by which carpenters are generally known, reflects this variety of roles. Daiku translates literally as "great craftsman" or "chief builder." It is significant to recall that the classical Greek term "architect" actually also meant "chief builder." The Japanese master carpenter, therefore,

Fig. 6. Chief master builder (*tōryō*) directs building assembly. Detail of pl. 4, *Matsuzaki tenjin engi emaki*, scroll 6, 1311.

should be seen as an architect in the Greek meaning of the word because of his responsibility for design as well as execution.

Originally the term *daiku* had no specific reference to the worker of wood. It was reserved for the upper end of the hierarchy in all the major building professions. Hence the chief roof tiler was known as *kawara-daiku*. During the seventeenth and eighteenth centuries the usage changed, perhaps reflecting the dominance of the master carpenter at the building site, and *daiku* became the generic word for a carpenter. The chief master carpenter became known as the *tōryō*, a metonymic title which means literally "ridgepole," the highest placed beam in the structure of the building and, by inference, the top of the carpentry hierarchy.

Within traditional carpentry practice there was an extensive hierarchy based on the master-disciple relationship common to all hereditary trades, religious institutions and performing arts. The tōryō was the director of the building project, supervising both its design and execution and answerable only to the patron or client (fig. 6). In the wake of Western influence, this role has been allocated to the architect who designs but does not build.

Fig. 7. Gatehouse (Sanmon), Nanzen-ji, Kyoto, ca. 1620.

Daiku were organized into family workshops based on a system of hereditary transmission of skills and techniques. Different workshops specialized in different types and styles of building construction, not unlike different sects within the same religious faith. The most specialized categories were the *miya-daiku*, the carpenters responsible for temple and shrine architecture, and the *sukiya-daiku*, who specialized in the construction of "teahouse-style" buildings especially for aristocrats, warriors and wealthy merchants.

The high degree of differentiation in building practices between carpentry specializations led to diversification of the form and function of the tools they used. The miya-daiku devoted their energies to large-scale structures with a preponderance of round pillars surmounted by multiple-arm bracket sets supporting the eaves (fig. 7). These temple and shrine carpenters used sturdy cutting tools to fashion large structural members and needed sophisticated marking and measuring instruments to determine the all-important proportions of the building and the curvature of the roof. Major temples and shrines had their own hereditary carpenters, as well as members of other building professions, living in the vicinity of their main gateway and charged with the upkeep of the buildings generation after generation. This symbiotic relationship with religious institutions undoubtedly influenced the organization and character of the miya-daiku workshops. Separate schools of carpentry (*ryū*)

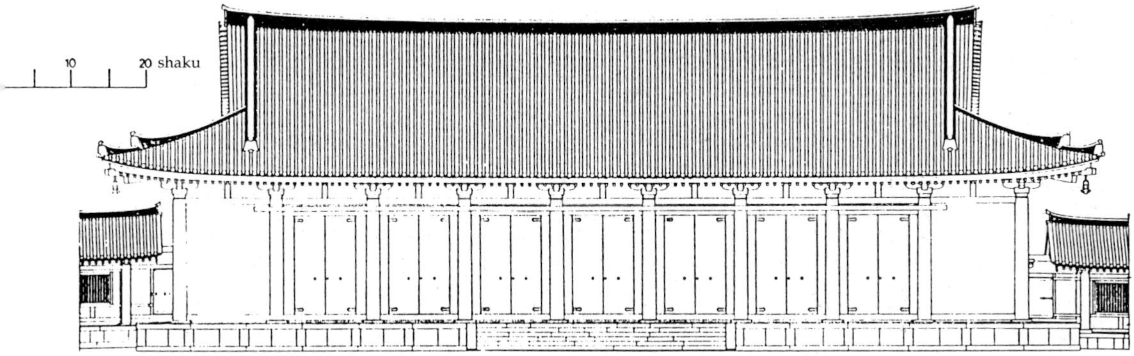

Fig. 8. Front elevation showing rectilinear style of temple design (Wa-yō): Great Lecture Hall (Daikōdō), Hōryū-ji, Nara. Late tenth century. *Hōryū-ji Restoration Report.*

Fig. 9. Front elevation showing curvilinear style of temple design (Zenshū-yō): Jizōdō (Hall to the Guardian Deity of Children), Shōfuku-ji, Tokyo. Late thirteenth century. *Shōfuku-ji Restoration Report.*

evolved from the experience of the workshops attached to great temples such as Hōryū-ji and Kennin-ji. The Hōryū-ji school of temple carpentry, based on the venerable center of Buddhist devotion and culture of that name near Nara, emphasized simple rectilinear design (fig. 8) while the Kennin-ji school, influenced by Song Chinese styles in the medieval period, delighted in complex curvilinear effects (fig. 9).

Fig. 10. Hamlet in Hakone Mountains, 1874. Farmhouse construction was the task of the entire community in premodern Japan. *The Far East*, 1874.

The sukiya-daiku was the magician with materials and interior finish rather than the master of large-scale structural framing. Although he used many of the same basic tools as did the miya-daiku, he worked with a wider variety of timber textures. The tateguya, for his part, specialized in making interior fittings, particularly delicate shoji and fusuma frames, as well as wooden doors and shelving. His tools include highly specialized hand tools similar to those used by cabinet makers in the West.

The *kobiki*, or sawyers, who cut logs into required lengths and thicknesses, were a crucial part of the process of constructing a building, though they came to be regarded as a separate profession from the daiku. They specialized in the use of logging saws and wide-bladed ripsaws for cutting planks, and some of these saws were lethal if misused.

Many specializations, therefore, fall within the broad category of carpentry. With the passage of time, these specializations became increasingly strict in definition, analogous to the emergence of trade distinctions among building guilds in Europe in late medieval times. A similar process of separation of tateguya from daiku (see chapter 4) may be seen in the distinction that emerged in England in the sixteenth century between the Guild of Carpenters, responsible for structural work, and the Guild of Joiners, responsible for doors, windows and paneling.

At the other end of the scale there were carpenters who were not members of

Fig. 11. Framework of a thatched roof farmhouse held together with rope lashings.

any specialist building trade. These were the part-time builders of the rural dwellings, the traditional farmhouses which have now all but disappeared from the Japanese landscape (fig. 10). Although assisted by a carpenter, who moved around the entire district determining floor plans and making the most important pillars and beams, most of the building process was in the hands of the people of the village. Men and women worked together in teams, performing the task of construction and periodic repair in the same spirit of community solidarity in which they planted and harvested rice. They used farming implements such as scythes and knives, along with rope and large wooden mallets, to assemble the structure. The framing on which the outer covering was arranged frequently consisted of little more than light saplings bound together with smoked vines or straw rope (fig. 11). Thatching was done with a large needle measuring up to a meter in length and a wooden paddle used to beat the straw or reed into place. In many regions roofs were covered with wooden shingles, which required little skill to place on the roof frame. They were simply laid out in regular rows and weighed down with a framework of bamboo and heavy stones.

This was building work far removed in tools and techniques from the sophisticated mortise-and-tenon joinery framing of the great temples and warrior mansions. It was an integral part of the way of life of the countryside but was not a part of the specialist Way of the Carpenter.

THE CARPENTER'S WOODS

The mode of selection and preparation of wood is the second key to understanding the Way of the Carpenter. Function and aesthetics dictated selection of materials for the various parts of the traditional building, and woods were chosen for their role according to their special structural properties and aesthetic attributes. In traditional practice, the master carpenter himself preferred to select the most important timber for a building from trees in the mountain forests. The dimensions and character of the wood available for a project helped to determine the form of the building. The natural curve of a beam might dictate the configuration of the entire truss. A prized piece of wood owned by the patron or a pillar taken from an earlier building might set the mood and design for a particular room.

Many different types of timbers are used in Japanese buildings. These can be divided into separate structural and decorative categories.

Structural Timbers

A timber-frame forms the structure of traditional architecture in Japan and softwoods have been the preferred timber for this role. "Softwood" is not necessarily a soft piece of wood but a misleading term for coniferous woods, many of which are quite hard. These are distinguished from broadleafed trees, which are known in general as "hardwood."

Hinoki, Japanese cypress (*Chamaecyparis obtusa*), is a softwood that is the preeminent structural wood used in Japan. It is a conifer notable for beauty, strength and resilience without undue weight or propensity to warping. It grows to a height of 40 to 50 meters with a fine, straight grain (fig. 12). These characteristics make it easy to work with hand tools and to split lengthwise with wedges to fashion pillars and beams. In modern times, hinoki grown in the cold, mountainous regions of Kiso (Nagano Prefecture), Tamba (Kyoto Prefecture) and Yoshino (Nara Prefecture) are the most valued timbers in Japan. It has been grown and cut at the foot of Mount Odaigahara, known as the "roof of the Kinki region," for many centuries. In ancient times, hinoki flourished in the mountains throughout the Asuka-Nara area, which was the focus of major building activity. It was used for most monumental buildings erected between the sixth and the thirteenth centuries. Hinoki pillars still survive in such buildings as the five-storied pagoda of Hōryū-ji, which was constructed over a thousand years ago.

One of the most remarkable features of hinoki is its longevity. Scientific tests have found that it is 30 percent stronger two hundred years after it is first cut. Thereafter it gradually weakens until a thousand years later, when it has returned to its original strength. The secret of its strength lies in the fine grain (only about one millimeter between growth rings) which aids oil retention.

Diminishing supplies of hinoki forced builders to substitute red pine (*akamatsu; Pinus densiforia*) and, more importantly, *keyaki* (*Zelkova accuminita*) for the sweet-tempered hinoki. In their natural state these woods grow curved or crooked, making it difficult to fashion them into the straight parts of a building.

Fig. 12. *Hinoki* (Japanese cypress) pillar, Main East Gatehouse (Higashi Ōte-mon), Nijō Castle, Kyoto. Early seventeenth century.

Fig. 13. *Keyaki* (zelkova) pillar, Belltower, Kiyomizu-dera, Kyoto. Early seventeenth century.

Moreover, red pine is vulnerable both to damp and to attack by insects. The deciduous keyaki, which closely resembles elm, is hard and has a convoluted grain that makes it difficult to work (fig. 13). Although stronger than hinoki when first cut, its coarse grain (up to five millimeters between growth rings) causes more rapid loss of essential oil; it dries and ages more quickly and is less able to resist rainwater as a consequence. It is not hard to imagine that such a radical change in timber type had a strong affect on carpentry tools and techniques. This will be examined more closely in chapter 4.

The use of hinoki, red pine and keyaki was limited principally to the monuments and residences of the elite because of high cost and status prerequisites. Black pine (*kuromatsu; Pinus thunbergi*) and horse chestnut (*tochi; Aesculus turbinata*) were favored by farmers and townspeople to build their residences and workplaces because of the lower price and ready availability.

Decorative Timbers

Connoisseurship in the selection and placement of woods required for the interior fittings determined the character of residential architecture, especially in the *sukiya*, or teahouse, style. Choice ranged from consciously rustic, unhewn

Fig. 14. Tokonoma with main pillar.

timbers, bark, thatch and bamboo to refined woods, such as the imported rosewood (*sendan; Melia azedaruch*) and ebony (*kokutan; Diospyros ebenum*) used for the ornamental shelving in the Shin Goten of Katsura Imperial Villa.

Japanese cedar (*sugi; Cryptomeria japonica*), a straight-grained conifer that is soft and easy to work, is prized for ornamental lintels (*nageshi*) and door panels. Kitayama sugi, artificially modified during growth to produce an irregularly patterned surface, is coveted for *tokobashira*, the main pillar framing one side of the tokonoma alcove (fig. 14). Unhewn Japanese cherry (*sakura; Prunus japonica*), with its natural textured surface and rich color, is also frequently found in this position. In sukiya architecture, many of the over one hundred species of bamboo available in Japan form lively tokonoma lintels and the eaves raftering adjacent to the garden.

The characteristic large burrs and lovely figured grain of keyaki, which make it difficult to work with hand tools, ironically make it particularly desirable for decorative features such as the main ledge of the tokonoma. The timber's hardness is well suited for carved pillar nosings, transom friezes and other sculpture. From 1668, sumptuary legislation issued by the Tokugawa shogunal government restricted the use of keyaki for gateways and cedar for door panels to the mansions of the highest ranking members of the samurai class.

Shinto shrines and many later Buddhist temples had raised wooden floors, frequently made of hinoki. Farmhouses and urban dwellings had one room with a floor of pounded earth (*doma*), sometimes stone flagged, while the raised wooden floor of the rest of the house was made of locally available timbers such as black pine.

In contemporary Japan, these structural and decorative timbers are used only for upper-class residences, restaurants and inns built in traditional style. The threatened exhaustion of indigenous forests forced importation of hinoki in large quantities from Taiwan in the postwar era. This timber has a lower density and wider grain than Japanese hinoki because it is grown in a warmer climate. It also has an alarming habit of splitting at an angle of forty degrees to its main axis. Moreover, it shrinks and changes color to an unattractive grey hue, very different from the constant dimensions and warm brown hue of native hinoki. Lauan (*Shorea contorta*) from the Philippines, Malaysia and Indonesia is used for Western-style interior finishing. Two-by-four studs imported from North America are now a common substitute for conventional Japanese timber framing and wall construction.

The character and qualities of the timbers used in traditional construction do much to account for the character of Japanese woodworking tools. Reliance on softer density conifers for structural purposes and some interior detailing has lead to the razor-sharp cutting edges on saws, chisels and planes for which Japanese tools are renowned. A wood of lower density demands a tool with a sharper blade if the cutting stroke is to be carried out cleanly.

It should also be remembered that many of the timbers are exceedingly valuable. In traditional practice the most important timbers were specially selected and often transported over great distances. The main pillars for the rebuilding of the great temple complex of Tōdai-ji in Nara at the end of the twelfth century had to be brought from mountain sides over a thousand kilometers away to the west. Today a single trunk of top-grade cedar may cost as much as an imported Mercedes-Benz. A carpenter cutting such expensive woods literally cannot afford to make a single mistake, another reason for insistence upon superlative quality in his tools.

METHODS OF BUILDING DESIGN AND CONSTRUCTION

Historically there are many different types of Japanese buildings and a variety of techniques for building them. Although most share common features such as a timber-frame structure with in-fill walls and a graceful overhanging roof, Buddhist temples and Shinto shrines, residences and castles, each required a specialized branch of the carpentry profession to build them. Each of these building types was different in scale, proportions, site organization, decorative emphasis and symbolism. There are cinnabar-red painted temples with massive framing like Tōdai-ji and delicately proportioned plain wooden shrines like Ise. Sumptuous palatial complexes like Nijō Castle in Kyoto stand in sharp contrast to the restraint of Zen-inspired halls of meditation at the Ryōgen-in of Daitoku-ji two kilometers to the north. The daunting scale of the walls and keep of Himeji Castle is difficult to equate with the microcosm of the three-and-three-quarters tatami-mat tearoom which became popular at the same time as the fortifications of Himeji were constructed at the turn of the seventeenth century. Dramatic differences in form and appearance do not always correspond with differences in function. Shrines are usually restrained and temples ornate in appearance. However, one of the most famous buildings in Japan is the Nikko Tōshō-gū, a shrine built with an ostentatious enthusiasm exceeding that of even the most exuberant temple or palace. It contrasts sharply with some temples displaying Shinto-like restraint such as Murō-ji, south of Nara.

The desires of the patrons, the conventions of the carpenters and other craftsmen, the nature of the site and the character of the available materials, together dictated the form of the building as these factors also dictated the specialist workshop to which the patron turned. Yet all buildings shared a common sequence to design and construction.

Siting

After consultation concerning the size, function and budget of a building, the master builder examined the site to determine the building's exact placement and favorable orientation based on both practical and geomantic considerations. A southern orientation was the almost universal rule where conditions permitted, for it assured both maximum sunlight and protection from malevolent cosmic forces dwelling in the north.

Measurement and Customary Design by Modules

In traditional practice, the construction of the building was not based on detailed architectural drawings determined by stylistic theories but upon modules grounded in traditional practice. Architectural drawings came into use only by the sixteenth century, but more as a guide for patrons than for directing the builders' activities. The basic organization of the construction process was based on a simple grid plan showing pillar placement set out on a large, flat plan board kept at the building site.

Many of the precise dimensions of the parts of a building were more the natural result of a process of interaction between man, materials and tools than of arbitrary design will. Architecture was the art of the possible, and the measurements of early buildings, including the famous temples and palaces of Nara and Kyoto, were determined as much by the way the pieces of wood split or could be cleaved as by any conscious intent.

Detailed designing was carried out during the construction process using proportions based on modules that were themselves relative to the circumstances of practice. These proportions, or *kiwari* ("wood dividing"), were applied automatically, and in many instances subconsciously, by the builder in accordance with custom. Systems of proportions differed according to period, region and individual workshop but were basically similar to those operating in Classical and Renaissance architecture. In Japan, as in the Mediterranean region, the primary module was the span of the post-and-lintel bay, the *ken*. This was divided into six subunits called *shaku* ("feet"), each further divided into ten *sun* ("inches"). Each *sun* consisted of ten *bu*, and each *bu* of ten *rin*. Thus a pillar might be one *shaku* wide, or one-sixth of the width of the bay, in a number of buildings but the actual length might differ according to workshop tradition, function of the particular building, or size and type of available structural timbers.

The tatami mat became a second important module of design in the sixteenth and seventeenth centuries. With floors completely covered in tatami, pillars were positioned according to the dimensions and arrangement of mats, reversing the previous procedure in which mats were lain between existing pillars. This marked a shift from a structural module to a spatial module in Japanese design. Like the *ken*, however, dimensions were not standardized until the emergence of three basic mat sizes by the eighteenth century. Of these the *kyōma* (1,970 millimeters by 985 millimeters), which was used in Kyoto, was the largest. The *inakama* (1,880 millimeters by 909 millimeters) was mainly found in the Kanto region, while the *Edo-tatami* (1,757 millimeters by 879 millimeters) was associated with Edo, the city that was the headquarters of the Tokugawa shogunate from the seventeenth century.

These measurements and modules also governed the dimensions and proportions of tools. Today the international metric system is in universal use in Japan. In 1886 the Japanese government standardized measuring units according to metric values in order to simplify international trading relationships. Traditional modules thus came to have the specific metric measurements of a 3. 03 centimeter *sun*, a 30.3 centimeter *shaku* and a 1.82 meter *ken*.

Many of the dimensions of tools used today are still determined by the traditional units as standardized in 1886 and, often as not, are based on the simple logic of the older rule of thumb. Expressing such measurements in modern metric equivalents tends to disguise their logic. For instance, it seems puzzling that a modern carpenter should have a jack plane which is 60.6 centimeters in length until it is remembered that under the traditional measurement system this was exactly 2 *shaku* long. Moreover, the width of the arm of the carpenter's square, the blade of a medium sized butt chisel and the sole of a plough plane all measure 1.515 centimeters—not the most meaningful of modern measures—which corresponds to 5 *bu* in traditional measure, 5 *bu* being one of the units most commonly used as a module in interior details of Japanese buildings. Measurements used for Japanese tools, therefore, make better sense if the traditional units are kept in mind. For a fuller understanding of the dimensions of Japanese tools, the table on page 8 provides traditional measuring units alongside modern metric and U. S. equivalents.

Selection and Orientation of Timber

The carpenter believed that the spirit of the tree had to be accommodated in its new location at the building site, but he had to know the environment in which it grew to maturity for practical reasons as well. In this context the old adage that the carpenter "does not buy a piece of wood but buys the mountain" (*ki o kawazu yama o kau*) is very true. The original orientation of the timber to the north or south had to be respected. This was not only to placate its spirit but also

Fig. 15. *Top,* a carpenter using a chisel and mallet to cleave a piece of straight-grained timber; *center,* a snap-line to mark a beam; *bottom,* and a *sashigane* and bamboo brush to draw a line on the end of a pillar. Detail of pl. 2, *Kasuga gongen genki-e,* scroll 1, 1309.

to ensure the suitability of a timber to its new location in a building; south-side wood is better adapted to warmth and should once more face south, whereas north-side wood is better able to withstand cold temperatures as it confronts the penetrating winds of winter.

Prefabrication of the Parts

Major pillars and beams and much of the interior detailing were prefabricated in the carpentry workshop before assembly at the building site (pl. 2). Each timber was measured and marked for cutting and labeled to indicate its location on the grid of the plan board. The ink pot and snap line (*sumitsubo*) and carpenter's square (*sashigane*) were the main tools in this process (fig. 15).

The most time-consuming part of the prefabrication process was the cutting of the mortise-and-tenon joints which held the building together. There were two main types of joinery: spliced or scarf joints (*tsugite*) (fig. 16) for connecting beams and sills end to end; and angled joints (*shiguchi*) (fig. 17) for securing members at oblique and right angles. Some joints, like the dovetail, were simple in design. Others were more complicated, particularly the larger horizontal

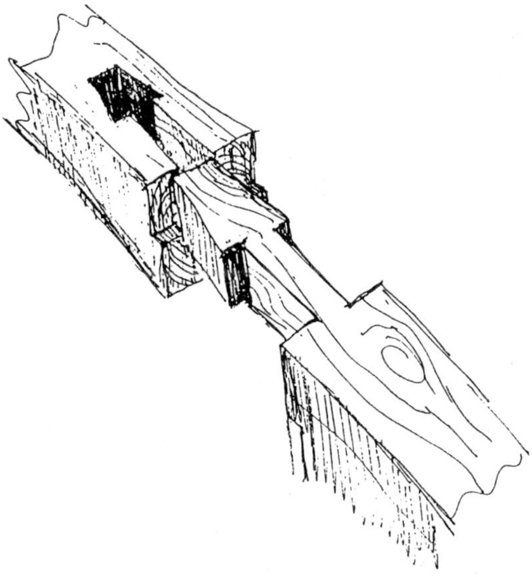

Fig. 16. Gooseneck mortise-and-tenon joint (*kama-tsugi*). Example of spliced or scarf joinery (*tsugite*).

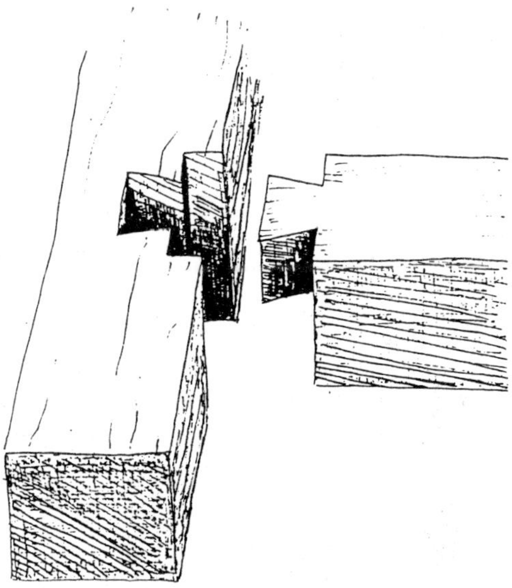

Fig. 17. Angled joint (*shiguchi*).

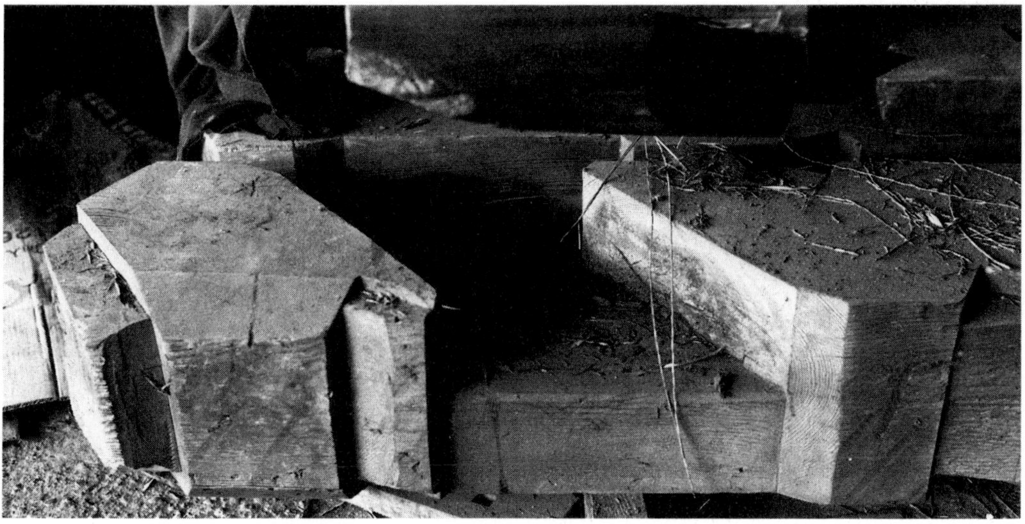

Fig. 18. Roof framing crossbeam with tenon and scarf mortise shown during restoration dismantling. Eighteenth century.

framing members with multiple beam intersections (fig. 18). Irrespective of form, absolute precision in the measuring and cutting processes was essential to ensure success in the assembly of the building.

On-site Assembly

After the ground-breaking ceremony, the site of the building was laid out with a grid of string markers corresponding to the plan board grid, and stone foundations were then positioned. Water levels were used to ensure that the ground or the podium on which the building sat had an horizontal surface. The prefabricated parts for the frame were then assembled on the foundations. The erection of the frame, which was a carefully rehearsed performance, took as little as a single day in the case of smaller residential buildings. Pillars were set on their footings and braced with horizontal beams and ties. Pent-roofed ("lean-to") sections were added around the completed core frame (*moya*) of the building. The roof truss of heavy horizontal ties and smaller vertical posts was placed on top of the wall frame (pls. 3 and 4). Even today the activity at a construction site during the assembly phase resembles an agitated beehive, with the carpenters swarming everywhere in rapid motion.

Completion of the framing operation was marked by the ridge-raising ceremony. Then followed the laying of roofing materials, which required a high level of skill and precision. Tiles were laid in steady progression from ridge to eaves and from right to left along the curving roof surface, pantiles (concave) preceeding covertiles (convex) and the ridge sealed with capping tiles. In the case of heavy terracotta tiling, its insistent vertical load also anchored the frame to the foundations and strengthened the joinery interfaces against earthquake jolts. Shingle and thatch roofs were built up from eaves to ridge and secured with bamboo cleats and wooden battens. The roof provided convenient shelter for the craftsmen who completed the walls and interior detailing. Sliding screens and tatami, which were also prefabricated, slid readily into their prescribed positions. Landscaping of the site proceeded concurrently, affecting visual transitions and integration of the building with its environment. Customary religious ceremonies marked the end of the project.

Having established a broad overview of the building professions and the activities of the carpenter at the building site, it is now possible to examine in detail the actual tools used for this process.

CHAPTER THREE

The Carpenter's Set of Tools

The fourth key to understanding the Way of the Carpenter is through his tools. The tools—the dōgu—link the men, the materials and the methods in building practice to create the structures and timber detailing Japanese architecture. What, then, are these tools and how are they used?

The simple answer is that Japanese carpenters' tools fall into four categories: (1) design tools, particularly marking and measuring tools such as the carpenter's square; (2) cutting tools, particularly saws and chisels; (3) finishing tools such as planes; and (4) tools for maintaining the tools themselves, such as whetstones and files. There are, thus, both "instruments of the mind" and "tools of the hand," which reflects the dual vocation of the daiku as designer-builder.

Some difficult problems of selection are encountered in explaining these tools in detail. Tool types and usage have changed considerably during the course of the fifteen hundred years of Japanese building for which records are available. If the tools used today only are described here, many of the valuable discoveries and much of the excitement in the Way of the Carpenter will be missed. Another problem is that frequently there is more than one explanation for how a particular chisel or a certain type of plane is used. There are also striking regional differences in Japanese carpentry techniques and tool usage. In premodern days a carpenter from the cold climate of northern Japan would have found much interest in comparing his tools with those of a carpenter from the warm regions of southern Kyushu—had they been able to understand each other's dialect. Even the names of tools and their parts may differ from one area to another and from one historical period to the next. There may also be some marked differences in tools used by carpenters specializing in different parts of the same building.

The tools a carpenter uses today are by no means identical with those used by his father a generation ago, and his grandfather would have been amazed at the changes that have taken place in the last hundred years. Even the basic form of such seemingly immutable tools as saws and planes has changed quite dramatically. An encyclopedia could scarcely contain sufficient space for the answers to the deceptively simple "what" and "how" of tools.

Most of the major tool types used in traditional carpentry are conveniently included in a 1870 photograph showing two daiku at work (fig. 19). In the foreground a carpenter uses a wide-bladed striking chisel (*tataki-nomi*) and steel-headed hammer (*gennō*) to start fashioning a mortise in a beam. Beside him rests a sumitsubo, the indispensable marking instrument for all carpenters. Propped against a wooden support is an adze (*chōna*), used for rough finishing of timber surfaces, together with a sashigane and a single-edged ripsaw (*tatebiki-noko*), indicated by the slightly concave curve of the top edge of its blade. In the background another daiku uses a finishing plane to create a smooth surface on a plank. Some caution must be exercised in assessing this seemingly irrefutable documentary evidence. The tools are real but a suspicion lingers that the men are not carpenters but actors set up by a keen foreign photographer of the exotic. What daiku ever worked on a beam lying directly on the ground without wooden blocks for stabilizing it and also helping to keep it clean? Even the angle of the planing board is too steep for actual use. Nevertheless, the information about different tool types which were in general use in 1870 remains useful.

Today the typical building site looks very different from this late nineteenth-century scene. The modern carpenter uses a combination of power tools and some hand tools for final finishing (fig. 20). Tools that were once mainstays of the building professions, such as the adze and the wide-bladed ripsaw, are now virtually extinct, and the number as well as the variety of specialized planes and chisels has decreased dramatically. However, standard tools in common use until a generation ago may be identified by checking the listings in a survey conducted by the Institute of Labor Science (Rōdō Kagaku Kenkyūjo) in 1943–44.[1] The survey reached the startling conclusion that the daiku needed a minimum of 179 tools for full-scale work. Most numerous were the chisels of which no fewer than 49 were deemed essential, reflecting their important role in making the mortises and tenons, the basis of the traditional structural system. Closely following the chisels in importance were the 40 required planes; 26 gimlets; 17 different measuring and marking tools such as squares, sumitsubo and marking gauges; 12 saws; 9 different nail pincers and claws; 6 hammers and mallets; 1 adze; 1 hewing axe; 6 whetstones; and 5 files. Western influence is also evident in the inclusion of a brace and bits, spirit level, adjustable wrench, spanner and 2 screwdrivers.

By the time the Institute of Labor Science survey was actually published, delayed in the aftermath of the war until 1959, its editors indicated postwar trends by noting that "today, with the gradual mechanization of building processes, there are some tools among these which can no longer be considered essential."[2] Karai Kunio of the Takenaka Carpentry Museum estimates that the ordinary carpenter today regularly uses no more than about eighty tools, including power tools. A top-quality carpenter may still possess over two hundred tools but will actually use only about one hundred and twenty of these.

Whatever the changes reflected in, or anticipated by, the Labor Science list of tools, it serves as an important benchmark for establishing the identity of tools in service at the end of the age of traditional building practice in Japan. This

Fig. 19. "Japanese carpenters" and their tools. *The Far East*, 1870.

Fig. 20. Hand and power saws at a modern building site.

chapter introduces and explains the characteristics and function of the most important of the tools in the listing, together with other essential instruments of traditional practice, which were disappearing even in 1943. Such doomed items include the long measuring rod, the plan board, manuals of building proportions and style, as well as scale models of buildings, all indispensable to the carpenter's role as both architect and builder in traditional practice.

Tools are presented, as far as practical, in the sequence in which they have been used at the building site. We shall start with design, measuring and marking tools, follow by cutting and finishing tools and conclude with tools used to make and maintain the others.

INSTRUMENTS OF THE MIND—TOOLS OF DESIGN

The master carpenter planned and prepared building construction with the plan board and measuring and marking tools, including the carpenter's square, the long measuring rod and the inkpot and pen. These are instruments of the creative mind, conditioned by the collective inherited knowledge of a workshop tradition, as well as by the personal experience of the individual carpenter. The master carpenter's phenomenal memory was aided by secret manuals of building proportions and style as well as models to establish overall spatial effects and guide assembly.

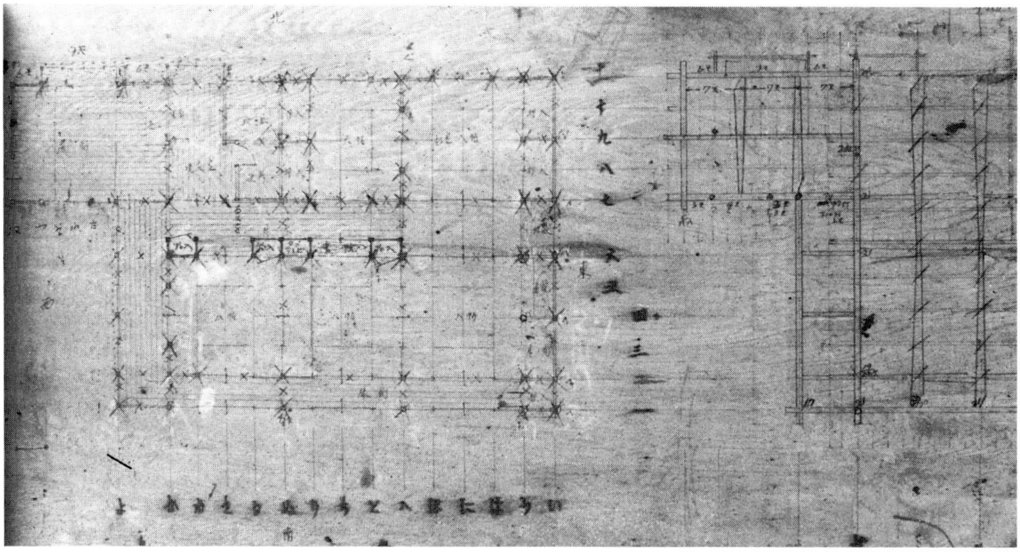

Fig. 21. Plan board (*ita-zu*), showing pillar placement.

Plan Board *(ita-zu* or *ezu-ita)*

At first glance, the plan board (fig. 21) may not appear to be a tool at all. We tend to see building plans as the province of the architect, separated from the process of building. However, in the Way of the Carpenter, the plan board serves as an instrument of cardinal importance in designing and building as one organic process. In traditional practice, the daiku as both the architect and builder drew the main drawings used for planning and constructing the building on this plan board.

The plan board bears little resemblance to the modern foreman's clipboard with its folded blueprints. As the name suggests, it was a simple plan set out on thin pieces of board. It varied in size but could be as large as three-by-two meters, depending on the scale of the project. A board was used for its durability and ease in handling. It could be propped up in a central position at the building site with the drawings clearly visible to all workers on the project.

The plan was boldly drawn in ink with the position of pillars, verandas and entrances marked. It rarely included any written notations or units of measure because the design of the building was based on customary modules, such as the span of the bay, discussed earlier. Detailed application of the module to decide actual dimensions was predetermined by the tradition of the carpenter's workshop, hence the absence of specific measurements. Constant reference was made to the plan board when the individual members of the structure were being measured, marked for cutting and labeled with their ultimate location. The individual members were identified by a grid system of cross references using numbers and the traditional *i-ro-ha* phonetic sequence. The most common system has the kana syllabary running across the top of the plan from right to left, the standard arrangment of any sequence in Japanese writing, with numbers running down the right-hand side of the board. There are regional variations; in the Tohoku region of northeastern Honshu, for example, the numbers ascend the board from the bottom right. For the more complex buildings like large-scale temples, a gable elevation would sometimes be superimposed on the plan, or even an entire elevation drawn, but for the most part the entire process of construction was based upon the simple plan board. No identification of the slope of the roof, instructions concerning materials or joinery type, nor any detailed working drawings were normally included. A daiku was expected to select unerringly the type of joint required for each situation and the type of wood that should be used for the different parts of the building. Even at construction sites today, some carpenters still take the scaled plan, elevation and sections prepared by the architect and then draw a traditional plan board for their working reference.

The Carpenter's Square *(sashigane* or *kanejaku)*

The measuring and marking of timbers (*sumitsuke*) is the most critical part of a carpenter's work. The success of the building project depends upon the accuracy of this process even more than it does in the West because there is no allowable margin for correction once the rapid assembly process commences. The accuracy of the measurements, therefore, will be confirmed only when the building is completed, by which time it is too late to correct deviations. Measuring and marking is supervised by the chief master carpenter. The first and foremost of his tools is the sashigane.

The sashigane (fig. 22) looks very similar to the Western set square but in Japanese building practice this tool is more versatile and of far greater importance. It forges the link between the basic plan set out on the plan board and the construction process by providing a system for calculating actual dimensions, angles and curvature. In other words, it is the means by which theory is translated into practice. For this reason alone it might be called the most indispensable of all the carpenter's tools because it is an implement both of design and execution, reflecting the role of the master carpenter as designer through the actual process of building.

The sashigane is used first to set right angles and to measure and mark the timbers in the prefabrication phase of construction. Various scales of measurement are indicated on both front (*omoteme*) and back (*urame*) of each arm (fig. 22). The outer edges of the front arms have the basic units in either traditional measure, from *ri* to *shaku*, or contemporary metric units. The arms themselves are 5 *bu* or 1.515 centimeters in width, which makes them a "ready reckoner" for one of the most useful of the traditional modules used by carpenters. Before all measurements were standardized in 1886, the sashigane and long measuring rule were used to establish the precise correlation between the modular order and physical dimensions.

The sashigane is also used by the carpenter to set angles. By trigonometry the sashigane determines angles other than ninety degrees and can set the graceful line of the eaves (fig. 23). For example, one *shaku* of horizontal run to six *sun* of

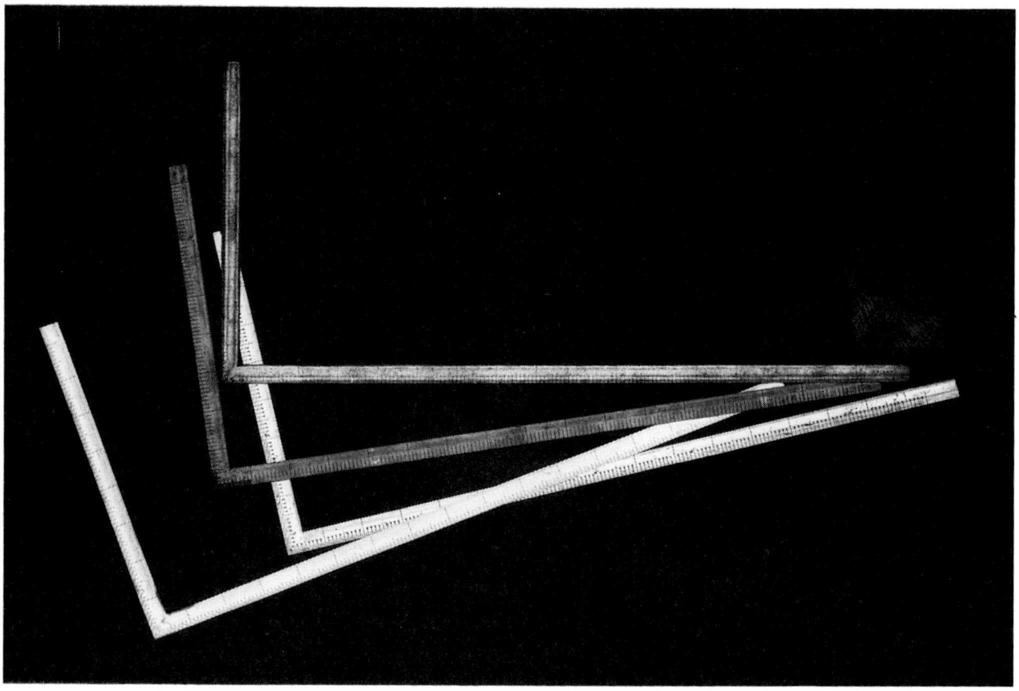

Fig. 22. Carpenter's squares (*sashigane*). They are made now from stainless steel but originally were fashioned from bamboo.

vertical rise creates a roof at the desirable rain-shedding angle of approximately thirty degrees (fig. 24).

The rear of the sashigane has several different scales that considerably simplify the daily work of the daiku as both designer and worker of wood, reflecting some of the special characteristics of the Japanese design tradition. Together with the units on the front, they furnish the specific dimensions, angles and curves set out as a general plan on the plan board. The rear arms are marked along the outer edges with the square root of the units of measure on the front side. This permits the carpenter to determine the curvature of roof-framing members by using tangents and parabolas based on the units of measure shown on the front scales. The most difficult part of the entire building to design and build was the eaves area, the *noki-mawari*, and many of the techniques for using the root scales on the rear of the sashigane were invented to ensure that the horizontal members and supports of the eaves edge rose in the precisely coordinated curves that are one of the most beautiful features of Japanese architecture. Building lore includes the delightfully evocative expression "swallows sing but carpenters weep at the eaves" (*daiku to suzume wa noki de naku*), which speaks vividly of the trials and tribulations involved in making sure that the two layers of rafters and their horizontal supports (*kioi* and *kayaoi*), employed for the most important buildings, were correctly positioned (fig. 30).

The root scale on the rear of the long arm may also be used to calculate the

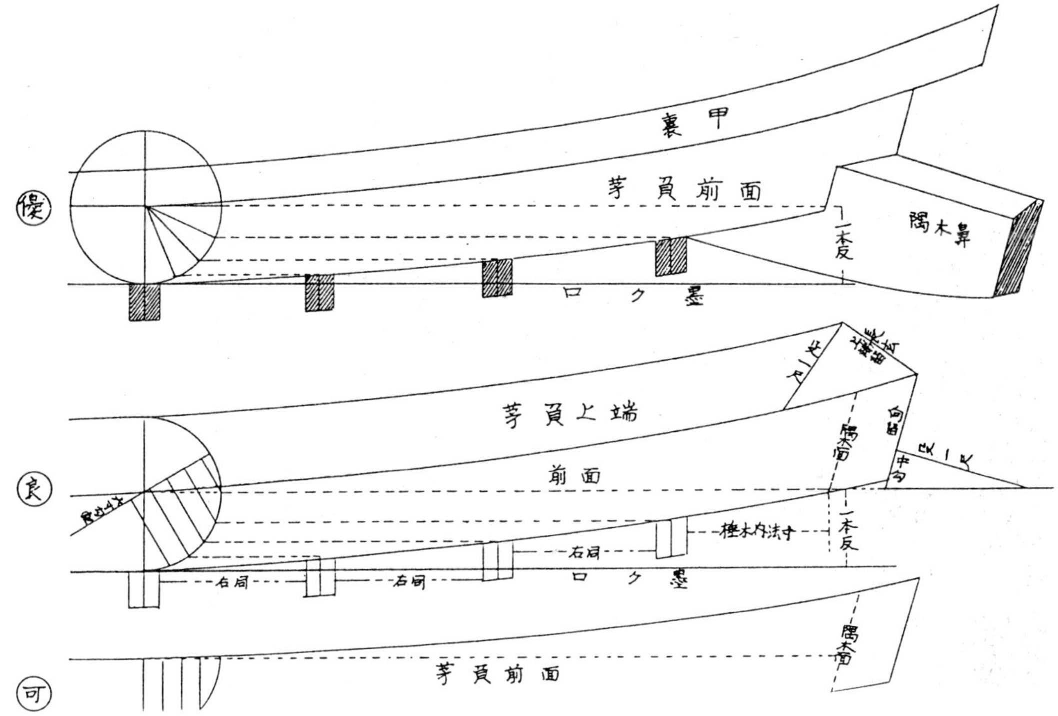

Fig. 23. Use of tangents with carpenter's square to determine the curve of the principal eaves beam (*kayaoi*).

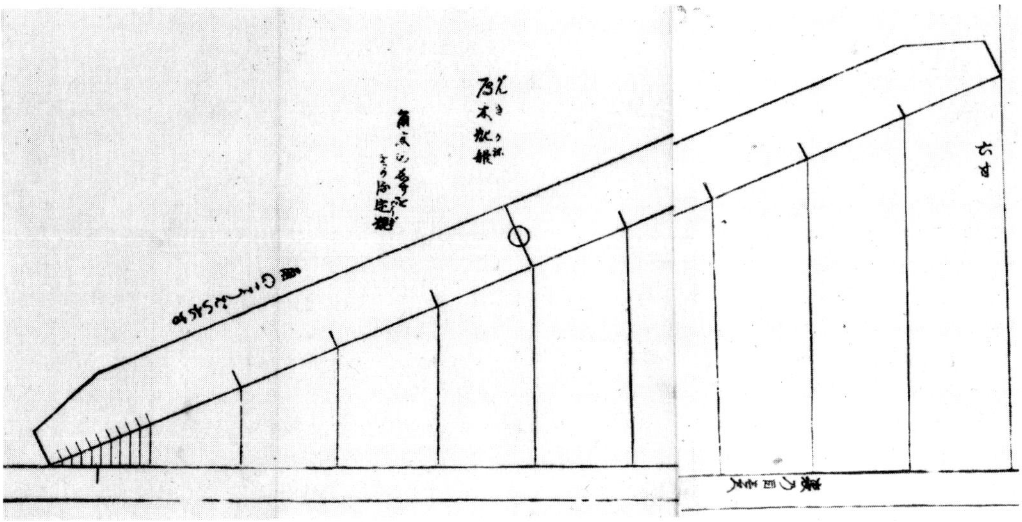

Fig. 24. Use of carpenter's square to determine slope of rafters ; in this case, 6 *sun* rise for each *shaku* of run. *Nokimawari taruki hinagata,* 1883.

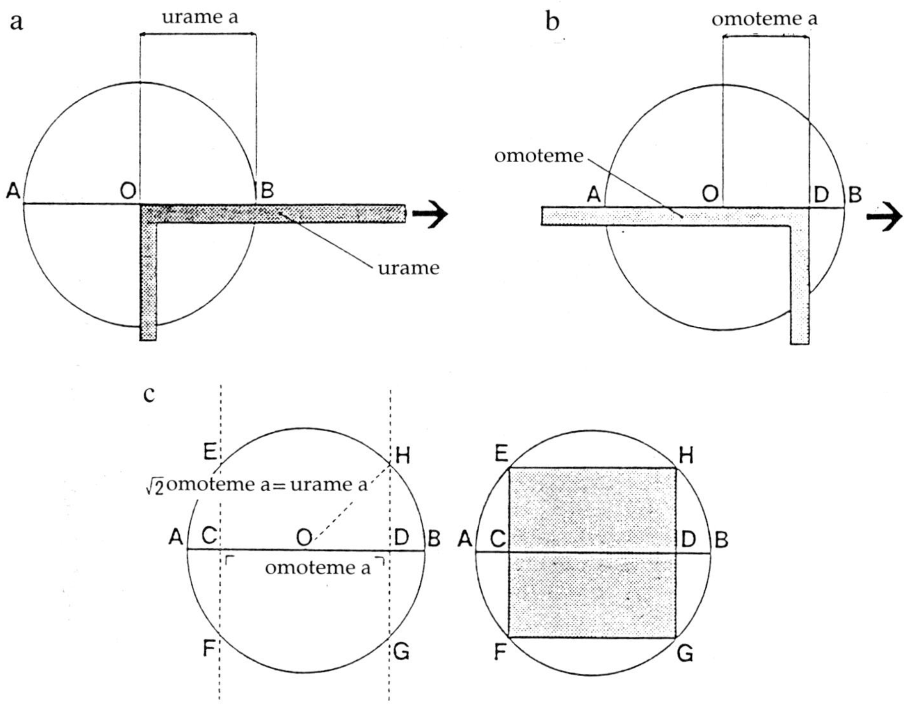

Fig. 25. Use of a rear square root scale (*urame)* on carpenter's square to determine the dimensions of a squared timber which can be cut from a log: (a) Draw the diameter AB. Measure the radius OB using the *urame* scale (*urame* a); (b) Flip the carpenter's square over to the front side (*omoteme*) and take a reading of the OB distance in standard measure (*omoteme* a); Use this measurement to determine the postion of D on the log; (c) Use the same procedure to fix the position of C. Then draw a straight line between C and D. Position E, F, G and H by drawing lines at right angles through C and D. Finally link up E, F, G and H to complete the dimensions of the timber which is to be cut from the log.

dimensions of a squared piece of timber cut from a round trunk. The ingenious procedure for performing this calculation is set out in figure 25, and its historical significance is explained in chapter 4. The rear of the longer arm also has a reversed standard scale, which is useful for determining restricted-access short measures like the depth of a mortise or the length of a tenon. In addition to these important functions of measuring and designing, the rear inside edge of the long arm has a special scale which divines good or ill fortune (*monjaku*), giving recognition to unseen forces in the lengths of the parts of the building.

The sashigane was once made from bamboo, which does not expand or contract to distort measurements and can be bent to draw curved lines. Today it is made from stainless steel with flexibility similar to bamboo. The arms are concave in section with the outer edges thicker to prevent the ink of the marking pen from blotting.

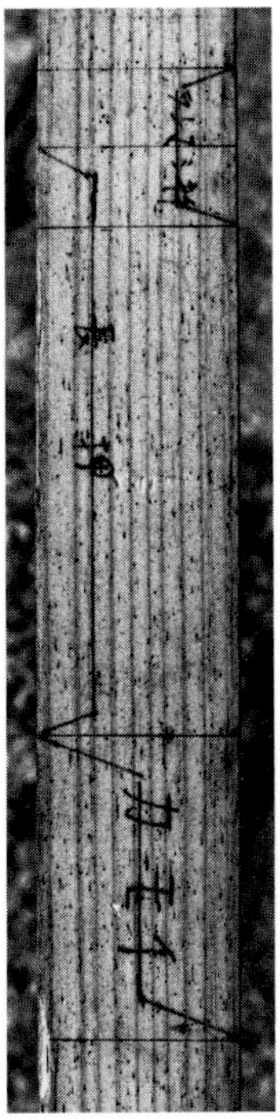

Fig. 26. *Left*, chief master builder holds long measuring rod (*kensao*) and plan board (*ita-zu*); *right*, markings on measuring rod to establish actual height of transom sill (*ramma shiki*), decorative sill (*nageshi*) and lintel (*kamoi*).

Long Measuring Rod *(shakuzue* or *kensao)*

The shakuzue is a long pole marked with *shaku* units. It was either 10 *shaku* (1 *jō*) or 6 *shaku* (1 *ken*) in length (fig. 26). This tool, along with the sashigane, was one of the simplest but most important instruments for turning the design set out on the plan board into an actual building. It established an exact linear value for plan board modules when laying out the building site and measuring the length of pillars and beams. As a major tool of both design and execution, like the sashigane, it was particularly important for designing on the vertical plane. The chief master carpenter marked the pole with the height of lintels, ceilings and other horizontal members after carefully studying the plan board. The exact measurements were based on his inherited knowledge of proportions. It was thereafter a simple and efficient procedure to mark off the position of these important members within the completed building frame.

The growth of a Western-style architectural profession in Japan from the latter part of the nineteenth century eroded the design prerogatives of the carpentry profession and rendered the practice of design by plan board, square and measuring rod obsolete, despite its efficiency and subtle accomplishments.

Carpenters' Manuals and Pattern Books

Carpenters' manuals and pattern books, known variously as *hidensho* ("secret hereditary writings"), *hinagata-bon* ("pattern books") and *gijutsusho* ("technique books"), are not "how to build" manuals in the sense of the trade manuals and amateur instruction books familiar to us today. They do not explain how to plane a beam or make a gooseneck mortise-and-tenon joint. They set out in written and diagramatic form the systems of proportions, or kiwari, which were the basis for designing buildings in traditional practice. They were based on the cumulative wisdom and experience of successive generations of carpenters in constructing buildings that were structurally sound and conformed to acceptable aesthetic norms. By the eighteenth and nineteenth centuries, the aesthetic considerations of proportions had become more important than structural concerns.

Although manuals date back to at least the fourteenth century, it was not until woodblocks were used to print them in the later seventeenth century that they came into general circulation outside the restricted family circles for which they were originally intended (fig. 28). This is discussed further in chapter 4.

These books contain written descriptions of the proportions to be employed in preparing the parts of the buildings, from primary pillars to lintels for the tokonoma alcove. Drawings accompanying the written texts set out the basic dimensions and show the method of using the carpenter's square for different building types and characteristic features, especially the roof gables. The manuals frequently consisted of separate volumes for gateways, *shoin*-style residences, temple halls and pagodas as well as shrine buildings. Vernacular housing was rarely included. However, the fact that the woodblock-printed manuals were available to ordinary town carpenters served to diffuse elements of high-culture style into the buildings of the general populace.

With the erosion of hereditary practice since the 1870s, these books have become an important aid to carpenters in preserving their building traditions. Contemporary publications continue to present information from the old manuals in updated form.

The *Buke hinagata* (Pattern book of samurai architecture), first published in 1655, is one of the most important of the manuals that establish standards for architectural design and offer insights into the approach to design in the Way of the Carpenter. The section describing the *munekado*, a gable-roof gateway used by important aristocrats and prelates (fig. 27), reveals the typical building blocks of the carpenter's design method. It is useful to see what it says in literal terms:

> The width of the [two] pillars should be calculated as a *sun*. The pillars are round. The principal crossbeams and the subsidiary crossbeams are 5 1/2 *bu* in height. Their lower surface [i.e., width] is one-third [of the diameter of the pillar].

From this statement we can deduce that the gateway is designed using two proportionally interrelated modules, the width of the bay and the diameter of

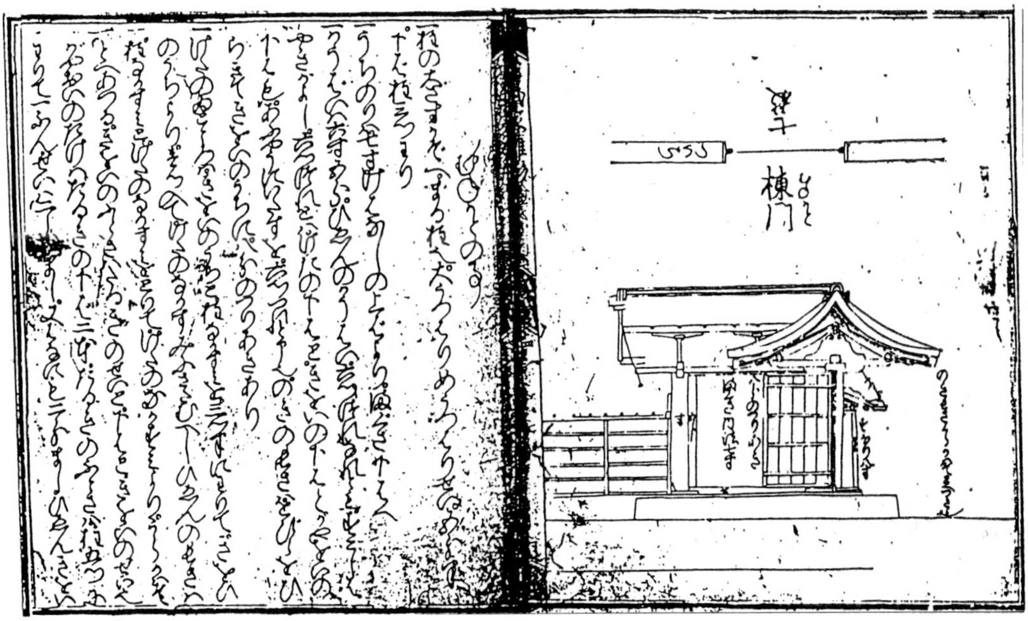

Fig. 27. Instructions for making a gabled roof gateway from "The pattern book of samurai architecture" *Buke hinagata*, 2d ed., 1658.

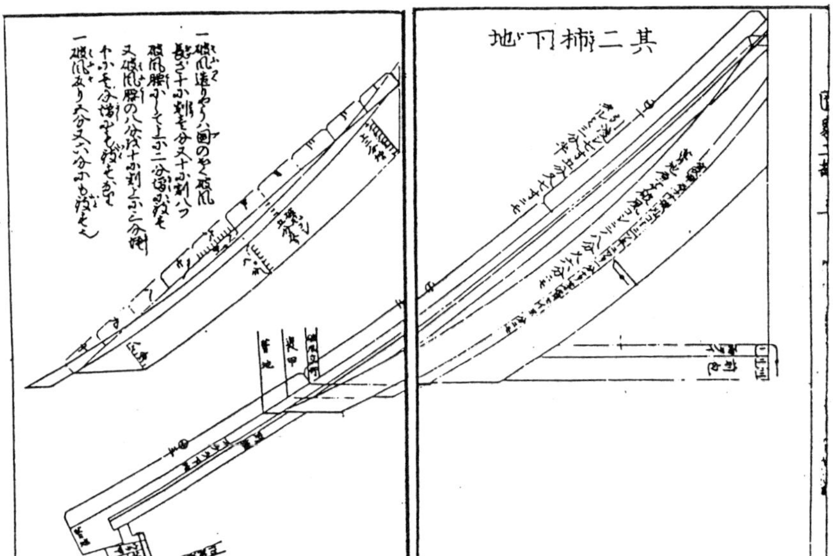

Fig. 28. Instructions in a carpenter's manual for determining the curvature of a roof gable. The instructions read: "To make a roof gable, divide the length of the gable into ten units (*bu*) as shown in the diagram. Divide the width of the barge board into eight subunits at the center but increase it to ten units at the inner end (edge of unit 1). . . . Drop the curve five units at [the edge of unit] 5. . . ." (*Shōke hinagata*, 2d ed.,1856).

the main pillars. We know that, in accordance with general practice, the actual measurements varied according to the particular structure. The terms *sun* and *bu* are familiar modular building blocks. The space between the pillars is considered to be one *shaku* module wide. As ten *sun* make one *shaku*, the pillar is one-tenth the width of the bay. Hence the carpenter is really saying: "The circumference of the [two] pillars should to be calculated as one-tenth of the width of the bay. The height of the principal and subsidiary crossbeams is 0.55 of the diameter of the main pillars, and their width is 0.34 of the diameter of the pillars." The similarity of these Japanese design manuals to the treatises of building proportions by Vitruvius and Palladio is intriguing, perhaps the inevitable result of a similar conjunction of trabeated building practice with the need for standardization within an increasingly idealized conception of style.

In the latter nineteenth century, some printed manuals were written to explain the complicated techniques for building the eaves using the sashigane. Until then, manuals had concentrated on aspects of building proportion, but the breakdown of the traditional training system for apprentices meant that basic building techniques had to be explained in written and diagramatic form for the first time. A good example is an 1883 manual that includes a diagram showing the intersection of the eaves horizontal support beam (*kayaoi*) with the corner rafter. This is the most important part of the structure of the roof frame and is critical for creating a graceful roof line (fig. 29). The manual also includes detailed diagrams showing the method for arranging the horizontal supports for double-layer raftering (fig. 30). It is a comment on the extraordinary circumstances of the day, the breakdown of hereditary transmission of knowledge and technique, that explicit information about fundamental building practices was freely distributed in printed form.

Models

Two types of models were used by daiku, wooden models and "pop-up plans" (*okoshi-ezu*). Both were employed in particularly complex projects to enable the daiku to judge the precise effects created by the combinations of materials, light and space, and to experiment with special design and assembly techniques. The pop-up plans were made from a single piece of paper and folded up to create a three-dimensional model (fig. 32). The wooden models were made to scale and showed major structural features as well as special details. A scale of one to ten was most frequently employed, which meant that models of larger buildings sometimes exceeded three meters in height (fig. 31). Model construction demanded great skill, and considerable care was needed to make all the structural features to precise scale. Sometimes models still remain for buildings that no longer survive, which makes them priceless historical sources.

Models are still made as part of the restoration of historic buildings, and many of them are preserved in the Tokyo National Museum and the National Museum of Japanese History at Sakura, Chiba Prefecture. Today it costs almost as much to make a model of a historic building as it does to build a ordinary house.

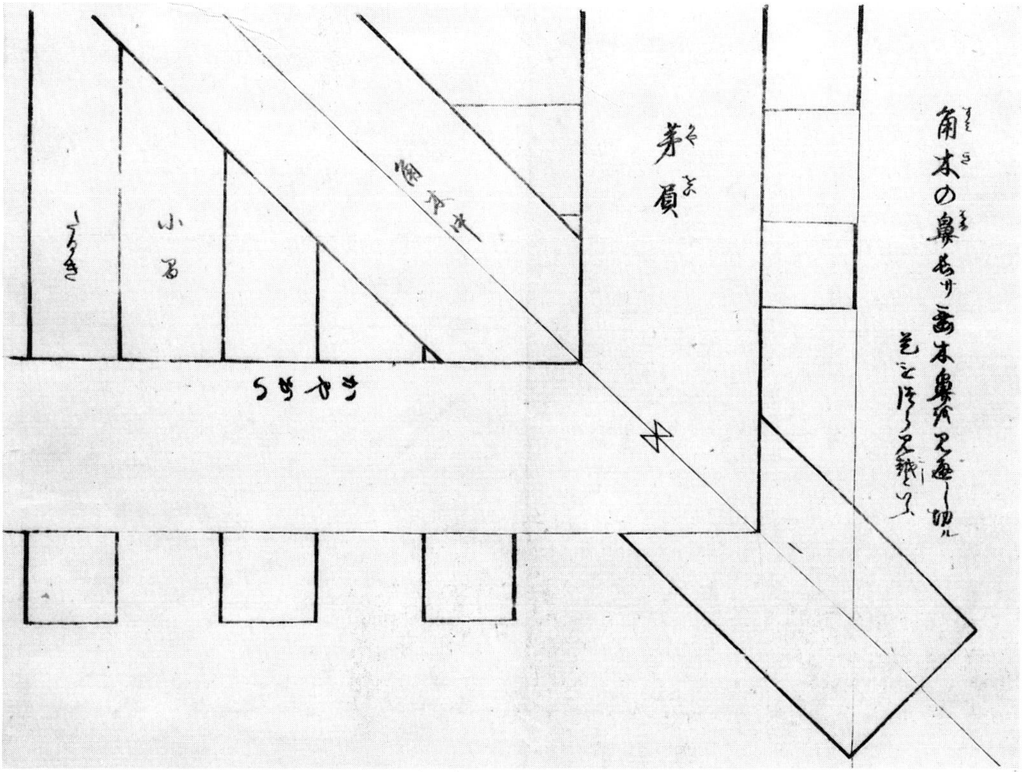

Fig. 29. Carpenter's manual showing correct method for making the intersection of two principal eave beams (*kayaoi*) with the roof corner rafter (*sumigi*). *Nokimawari taruki hinagata,* 1883.

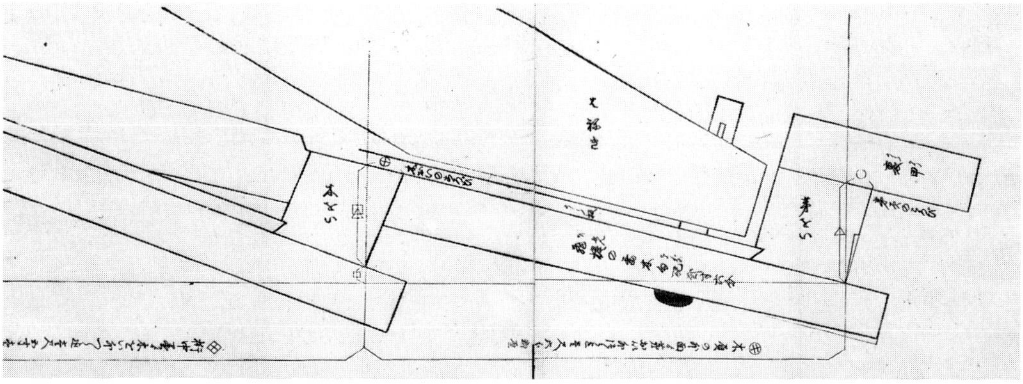

Fig. 30. Carpenter's manual showing construction of eaves with two tiers of rafters. *Nokimawari taruki hinagata,* 1883.

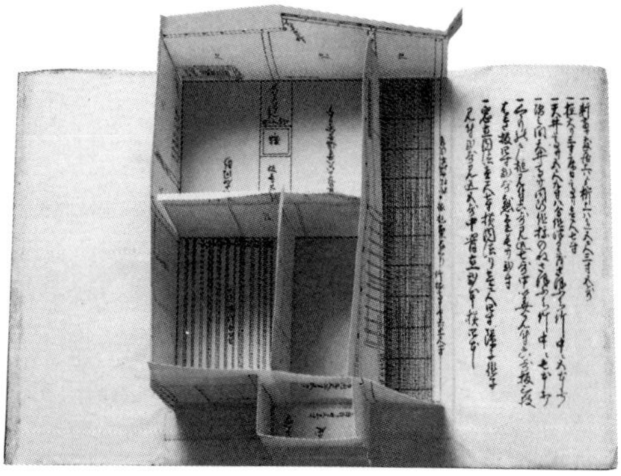

Fig. 32. "Pop-up" model of a teahouse.

Fig. 31. One-tenth scale model of East Pagoda, Yakushi-ji, Nara, made by a master carpenter. *Kintetsu Historical Museum, Nara.*

Ink Pot and Snap Line *(sumitsubo)*

The uniquely shaped sumitsubo (pl. 1, figs. 33 and 34) is the traditional symbol of the carpentry profession. Sumitsubo, richly lacquered and inlaid with mother of pearl, have long served with ornamental carpenter's squares and adzes at Shinto offering altars erected at ground-breaking and ridge-raising ceremonies (fig. 5). In appearance the sumitsubo resembles a shallow-draught river barge propelled by a paddle wheel. As its name implies, it is the standard container for soot-black ink (*sumi*) used in conjunction with a bamboo brush for marking purposes. It also serves a variety of other important functions including that of a snapline for marking timbers and a plumbline for ensuring verticality.

The sumitsubo consists of a block of wood, usually the intricately grained keyaki (zelkova), with one end round like a gourd and hollowed out to form the ink pot. A wad of cotton saturated in sumi is placed in this cavity. The opposite end holds a wheel-shaped spool wound with some seven meters of line. This line runs through the body of the sumitsubo across the inked cotton wad and out through a small hole drilled in the wood. The string is tied to a small piece of rounded wood that has a sharp pin projecting from it. The tool is now ready for marking long timbers with a straight line by the simple expedient of drawing

Fig. 33. *Sumitsubo* (ink pot and snap line), symbol of the carpenter.

the line out over the ink to a predetermined point on the timber. It is secured in position by pushing the attached pin into the wood. The line is then pulled taut and allowed to snap back against the wood. A clear, straight mark is left down the entire length of the timber, even when its surface is irregular. The string is wound again onto the wheel, thereby re-inking it in readiness for the next use. The line must be extremely light but strong, without any fine hairs or fluff which would leave a furry impression on the wood. The best type of line is made of thread from wild silkworms in Nagano and Hiroshima prefectures.

The sumitsubo works on the same principle as the chalk line of Western woodworking tradition. European sawyers also used a long string soaked with lampblack and water to leave a firm black line to guide their work.

The carpenter customarily has a separate sumitsubo for black and for red ink. The sumi is mixed with seaweed glue (*funori*) to make it insoluble in water so that important markings and numbering will not be washed away by rain during the building process. The red ink sumitsubo is intended for use in sukiya architecture, particularly on timbers in important parts of the building that will be visible to the naked eye. The ink in this case is made from soluble red ochre powder so it can be washed off after the measuring, cutting and assembling processes have been completed. In addition to its functions as an ink pot and pluck line, the sumitsubo is sometimes used as a plumb line to test verticality, its body acting as the plumb weight (fig. 6).

The size and decoration of sumitsubo vary considerably with the variation in the building practices of different workshops and regional locations. Most sumitsubo in use in residential projects are approximately 20 to 30 centimeters in length. Traditionally, larger ones were required for temple and shrine construction because longer pillars and beams were used. These were often twice the size of the ordinary sumitsubo and their line length was doubled to approximately 14.5 meters. Sumitsubo made in the Kanto region, around what is today Tokyo, have handles to roll up the inking line, but those from older

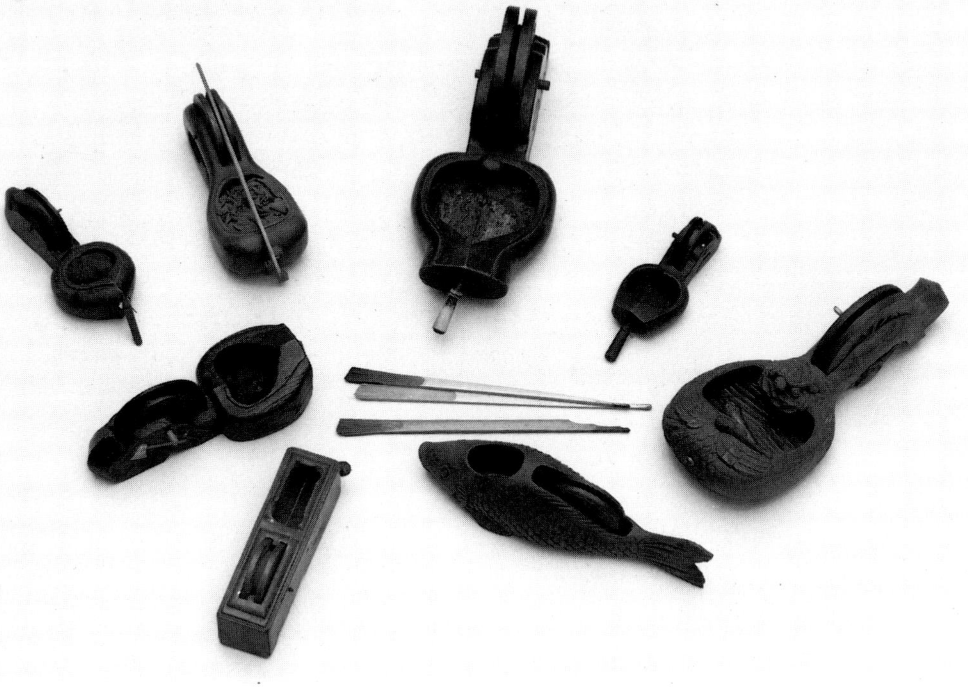

Fig. 34. *Sumitsubo* and bamboo drawing pens.

traditions in the Kansai (Kyoto, Nara, Osaka) region do not have handles, the line being rewound by simply rotating the wheel rapidly with a stroking action of the hand.

The ink pot provides the carpenter with the opportunity for self-expression through the carving of its shape as well as decoration. It is the most consciously decorative of all his tools, its surfaces often embellished with carvings of cranes, turtles and carp swimming in the sea. Decorating the body of the sumitsubo with felicitous creatures from the storehouse of Asian mythology became popular in the 1880s and 1890s with the establishment of a special profession of sumitsubo makers. At the same time, the shape of the ink pot became square with rounded corners. Earlier sumitsubo were shaped like a heart to represent the leaves of the bodhi tree under which Gautama Buddha is believed to have attained enlightenment. Thus the traditional design of the sumitsubo is another important reminder of the religious dimension to the Way of the Carpenter.

Bamboo Drawing Pen *(sumisashi)*

This tool (fig. 34) is one of the simplest in the carpenter's collection. It consists of a piece of bamboo usually 15 to 25 centimeters in length, although sometimes as

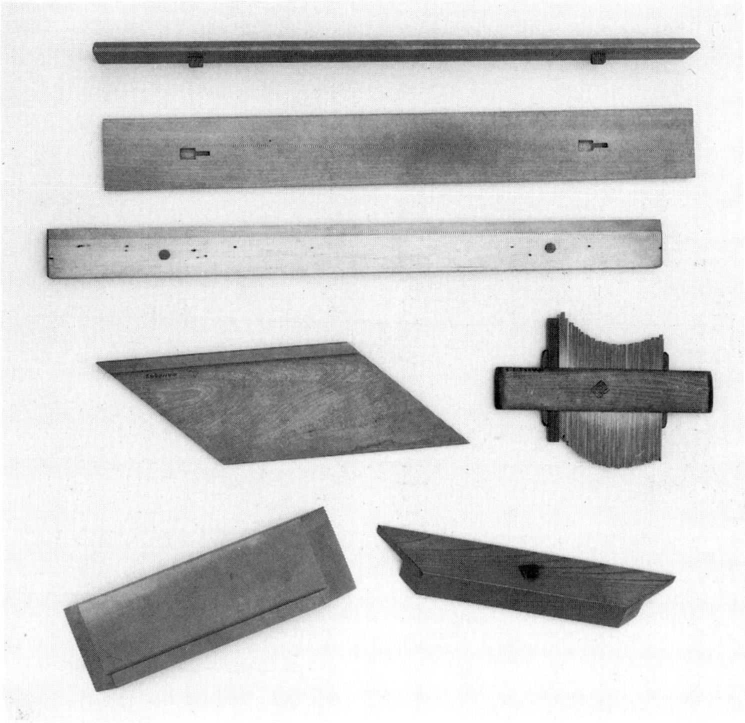

Fig. 35. *Top*, three plane sole gauges (*awase-jōgi*); *center left and bottom*, miter joint gauges (*tomegata-jōgi*); *center right*, pillar profile gauge (*osa-jōgi*).

long as 30 centimeters. Depending on the length, its width varies from 1.0 to 1.5 centimeters. One end is pointed, and the other is cut flat at an angle of about twenty degrees with approximately thirty lateral cuts 2 centimeters down the grain to form a coarse, brushlike bristle. The tool is used for marking and labeling timber to be cut by the simple method of dipping it into the ink pot. The pointed end is used for writing characters and numbers, and the flat, splayed end is used for drawing lines.

As the ends wear down with use, the bamboo inking pen is repeatedly recut until only a short stub remains, rather like a pencil. Moreover, in much the same way as the mechanical pencil with inserted lead has come into widespread use in recent years, a brass handle holding pieces of cut bamboo has been developed to replace the more humble inking pen.

Gauges

Measuring Gauges The two main measuring gauges (*jōgi*) (fig. 35) are the miter-joint gauge (*tomegata-jōgi*) for marking 45-degree angles and the block gauge (*hakogata-jōgi*) used to draw a line over two intersecting surfaces of a square pillar. A third gauge, the profile gauge (*osa-jōgi* or *osa-maruta-biki*), is a

Fig. 36. Technique for using the profile gauge (*osa-jōgi*) to transfer the irregular contour of the stone footing to the pillar base.

Fig. 37. Single line-marking gauge (*suji-kebiki*).

special measuring tool designed for transferring the irregularly shaped outline of a stone footing onto the bottom of a pillar (fig. 36). It consists of several hundred narrow strips of bamboo held securely in place by an iron frame. This is tapped over the surface of the stone, creating an accurate outline of the profile which is subsequently transferred to the bottom of the pillar.

Marking Gauges Marking gauges (*kebiki*) are used to place guidelines along the surfaces of wood. The marking gauges facilitate fast and accurate preparation of timbers to suit the most fastidious craftsman. They are especially important for marking timbers selected for interior fittings such as sills for sliding screens (*shikii*) and lintels (*kamoi* and *nageshi*). These guidelines must be laid out with great precision, particularly so that the interior detailing fits exactly and screens move smoothly.

Marking gauges are made from a rectangular wooden block with a narrrow outrigger-style arm set through it at right angles. Into this arm is set, in turn, the marking knife which lightly scores the surface of the wood. The gauge is adjusted by moving the arm, which is secured by a small wooden wedge or—with modern gauges—by a screw. The rectangular block itself runs along the edge of the wood to allow a straight, parallel line to be cut.

There are three principal marking gauges: (1) the single-line marking gauge

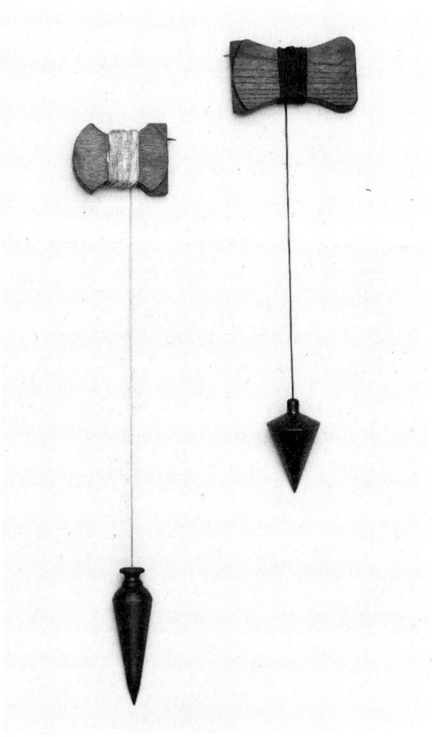

Fig. 38. Plumb lines (*sagefuri*).

(*suji-kebiki*), the basic marking gauge equipped with a single scoring knife (fig. 37); (2) the double-line marking gauge (*nimai-kebiki*), which has two knives set into the arm and is used to lay down the twin parallel lines required for cutting the grooves for sliding screens; and (3) the splitting gauge (*wari-kebiki*), which is used to score thinner pieces of wood deeply along the grain, thus permitting them to be broken cleanly into two pieces.

Plumb Line *(sagefuri)*

This tool consists of a heavy, cone-shaped bronze weight attached to a length of strong line that is wrapped around a wooden handle when not in use (fig. 38). A variation of the traditional plumb line, and now replaced by the Western spirit level, it was a simple upright braced by two struts with a weighted string suspended from the upright.

The Western spirit level is in general use both for determining horizontal and vertical planes and for checking flatness. The traditional instrument used prior to the introduction of the spirit level for checking horizontality was the *mizubakari* ("water calculator"). It consisted simply of a hollowed-out piece of timber into which water was ladled, the level of water indicating the horizontal plane.

TOOLS OF THE HAND—CUTTING AND FINISHING TOOLS

Chisels *(nomi)*

The chisel family is the largest single group of Japanese tools, reflecting its vital role in fashioning the intricate joinery holding the Japanese building together. The chisel has played a much more significant role in the evolution of Japanese architecture than occidental chisels played in the development of Western architecture.

The head of the chisel is made of carbon steel. This is forged onto a softer steel section to give it resilience and to conserve precious steel (fig. 40). The head is beveled on one side of the tip to form the cutting edge. Chisels with a head wider than 1.3 centimeters also have bevels down the shoulders of the blade. These bevels reduce resistance as the chisel cuts through the wood. The back of the head is also hollow ground or fluted to facilitate cutting. The head of wider chisels tapers into a funnel-shaped neck or shank, which gives it a distinctive appearance. This shank culminates in the tang, or spike, which is driven into a hardwood handle, usually made of oak or boxwood, and then secured by a steel collar. Thus, the chisel in Japan combines the socket and tang methods of attaching the metal blade to a wooden handle to give it greater strength than either the simple socket or tang method alone.

Like all Japanese cutting tools, the nomi is a thoroughbred. It is a precision instrument perfectly adapted to the tasks it performs but demanding, in turn, the highest level of skill from its user. The cutting blade is very hard, measuring as much as sixty-four on the Rockwell C scale of seventy, but as a result it is also quite brittle. The carpenter must exercise utmost care to prevent nicking the blade. The finer the steel, the greater the precision demanded of the user.

The precision of traditional Japanese chisel manufacture and use may seem excessive to us today. Modern glues have rendered precision unnecessary when making firm mortises and tenons. In the Japanese building tradition, however, a tight fit was not merely the ideal; it was the absolute minimum requirement for everyday practice. The combination of superb steel and painstaking honing at the workshop and building site resulted in a razor-sharp edge which was no mere craftsman's affectation. Such an edge was essential because of the prevalence in Japanese architecture of soft timbers, which require very sharp cutting tools.

Chisels were traditionally made of laminated steel of the unsurpassed excellence associated with the work of the master swordsmiths in the age of the samurai (pl. 9). A recent comparison of the performance of Japanese and Western-made chisels has revealed that chisel blades made by Japanese hand-forging techniques hold their cutting edge up to six times longer than those mass-produced by the Western drop-forged method.

We should not forget, however, that in part this qualitative difference is due to the greater emphasis on industrialized mass production of tools in the West. American chisels crafted before the industrial revolution by traditional hand-forging techniques were of superior quality than most of those manufactured

Fig. 39. Early seventeenth-century carvings of phoenixes, chrysanthemums and other flora and fauna under eaves of the Gateway (Kara-mon), Hōgon-ji, Chikubushima, Shiga Prefecture.

today. Perhaps the superior Japanese chisels of today would find more congenial company with a gouging chisel made in 1775 found by Eric Sloane in an old stone fence. "Bright and silverish, its edge is keen; it has no rust. . . . I have compared the best chisels (the most expensive, that is) by leaving them in the rain alongside this ancient tool. The new tool's edge was dulled, and rust appeared within a few days."[3]

The Japanese chisel is used for both structural and decorative work. It fashions the mortise and tenon joints, the key to the structural durability of the timber-frame building. Its use is equally important for carving the decorative fittings that embellish and animate the pillars and eaves of the great temples, shrines and palaces (fig. 39). There are as many different types of chisels as there are functions to fulfill. They range from powerful, wide-headed striking chisels driven by aggressive hammer strokes for cutting out large mortises, to the delicate, paring chisels used for sculpting the leaf-shaped nosings set at the top of pillars.

Chisels fall into two broad categories—striking chisels and paring chisels or slicks. There are almost infinite variations in dimension and shape within these two general divisions.

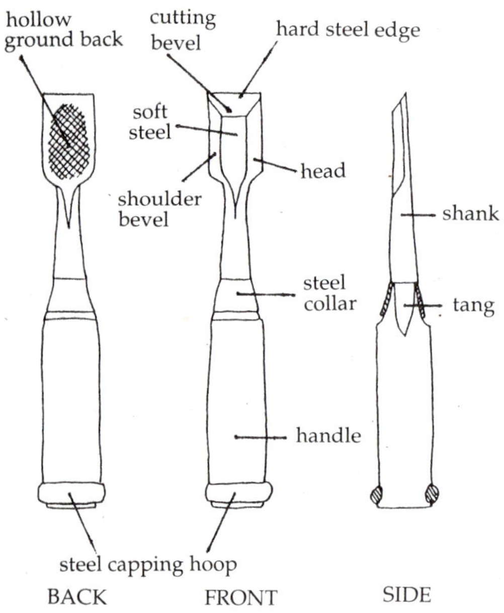

Fig. 40. Striking chisel nomenclature.

Fig. 41. Seated carpenter works with striking chisel.

Striking Chisel *(tataki-nomi)* The tataki-nomi is the most important type of chisel used by the carpenter in preparing the structural members and special interior fittings. Once in position, it is struck with a steel-headed hammer, the gennō. The end of the wooden handle is reinforced with a steel capping hoop to help cushion the forceful blows (fig. 40).

Striking chisels are used by the carpenter from a seated position (fig. 41). This enables him to hold the wood steady and allows maximum control of the operation of the chisel itself. The hammer blow must be delivered absolutely perpendicular to the axis of the chisel or the blade will not cut cleanly and the brittle head may be damaged.

There are two types of striking chisels: the framing chisels for making the mortise and tenons of the major structural timbers of the building, and the butt or finishing chisels for completing the interior finish and eaves detailing (fig. 42).

Framing chisel (fig. 43) The first chisels used by the carpenter in prefabricating a building are the sturdy striking chisels to cut the mortises. Mortise making is one of the most difficult tasks performed by a carpenter in traditional practice. Three years are considered necessary to master mortise-making skills. The Japanese mortise chisel, like its Western counterpart, has a blade shank 30 to 40 percent heavier than other chisels because it must withstand the insistent driving pressure of the sharp hammer blows needed to penetrate the densely grained cypress and zelkova woods used for framing timbers. The slope of the cutting edge is set at about thirty-five degrees, eight to ten degrees steeper than

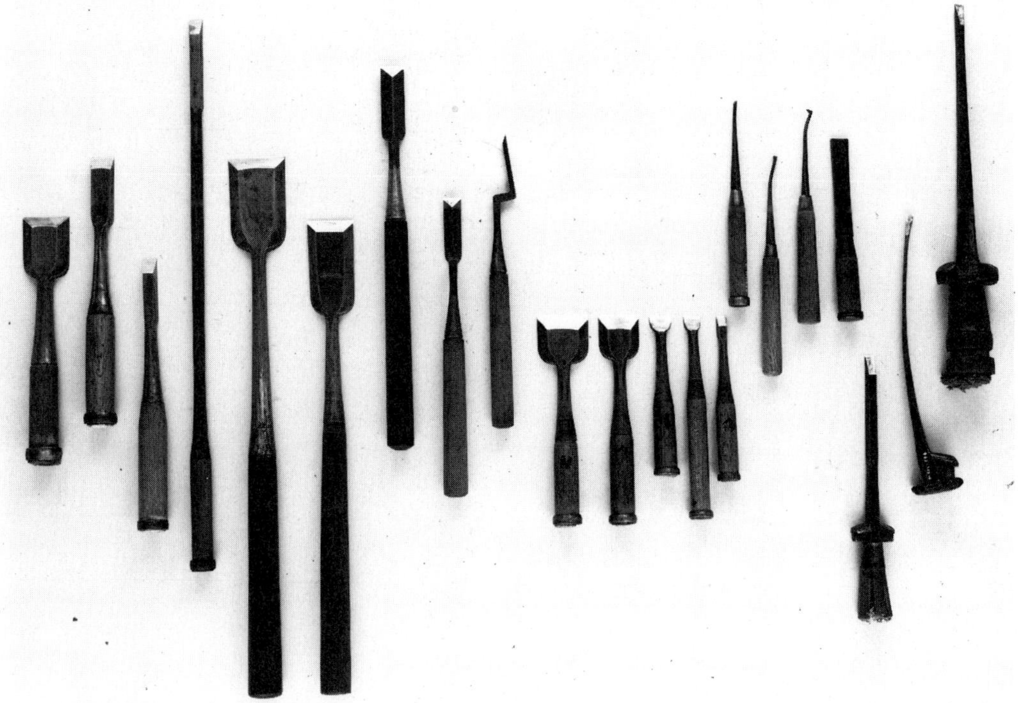

Fig. 42. *Left to right*, four striking chisels used for framing (*tataki-nomi*), two paring chisels or slicks (*tsuki-nomi*), two dovetail slicks (*shinogi-nomi*), one trowel chisel (*kote-nomi*), five butt chisels (*ōire-nomi*), three "bottom-cleaning" chisels (*sokosarai-nomi*), one "strike-through" chisel (*uchidashi-nomi*) and three "sword-guard" chisels (*tsuba-nomi*).

that of other chisels, to facilitate heavy-duty cutting. The shank of the chisel is at least six centimeters in length to enable the cutting of deep mortises. The requirements of different stages in the mortise-cutting process demand chisel heads of various widths. Those wider than about three centimeters (1 *sun*) in width, commonly called *hiro-nomi* ("wide chisels"), are used to outline the mortise with shallow vertical cuts, leaving the sumi markings still visible. The rough cutting stage is commenced with a medium-sized chisel, usually 2.5 centimeters (8 *bu*) in width (fig. 44). The hole is excavated by cutting at an angle of about forty-five degrees from each side alternately (fig. 45) once the medium and small-headed chisels have roughed out the hole. A wide chisel is then used to smooth the interior walls. Japanese master carpenters did not use gimlets or augers for drilling starter holes for mortises, a significant difference from the practice in the West.

A chisel with a shank of approximately 5.5 centimeters, fractionally shorter than that used for the mortise-cutting chisel, is used for fashioning tenons. Much of the work of making larger tenons is performed by using a saw, but more

Fig. 43. Framing chisels for cutting mortises and tenons.

Fig. 44. Starting a mortise with a framing chisel.

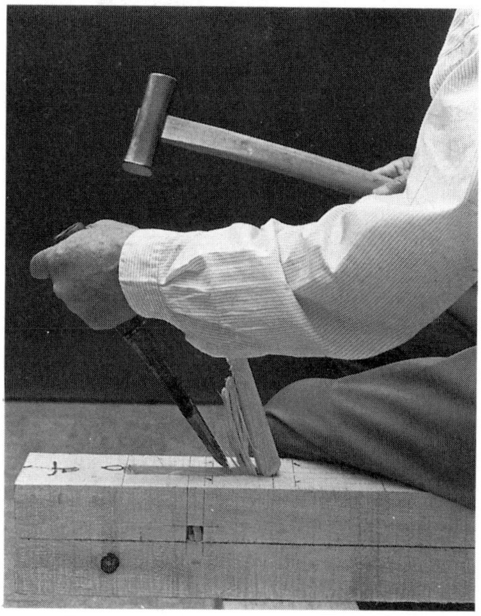

Fig. 45. Rough-cutting a mortise.

Fig. 46. Smoothing a tenon with wide-bladed striking chisel (*hiro-nomi*).

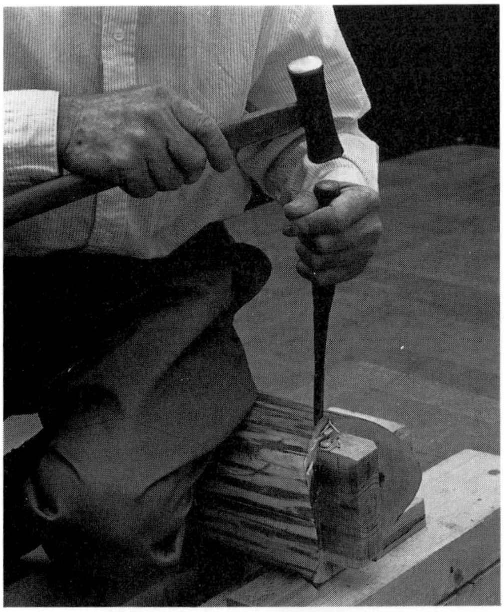

Fig. 47. Use of a butt chisel to make lintel/sill joinery.

Fig. 48. Set of butt chisels at building site.

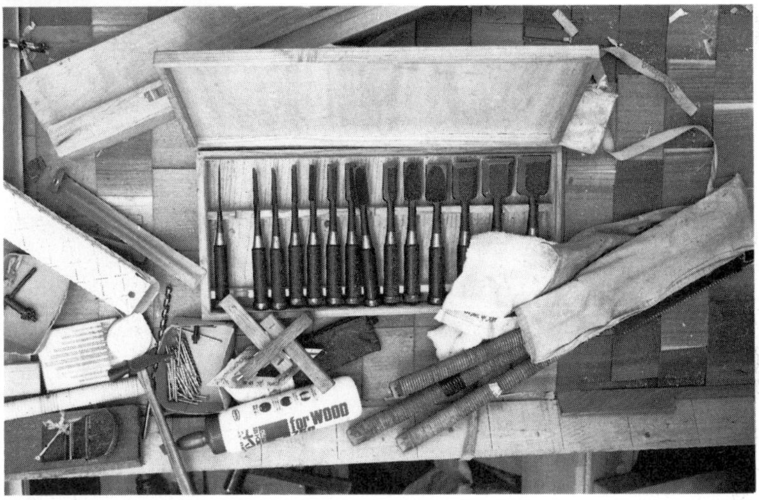

precise tongue edges may also be cut with this chisel. The short-shanked, wide-bladed striking chisel is particularly effective for smoothing the shoulder of tenons (fig. 46).

Butt chisel (ōire-nomi) Butt chisels are used for nonstructural work, particularly for the delicate joinery and bevels on lintels and sills (fig. 47). The shank of the butt chisel is about 4.5 centimeters long, 2 to 3 centimeters shorter than framing chisels in order to give more control in detailed work.

The master carpenter usually owns a set of fifteen butt chisels (fig. 48). The

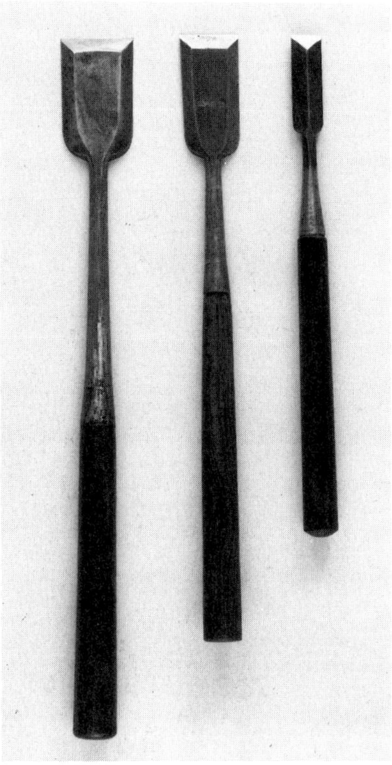

Fig. 49. Paring chisels (*tsuki-nomi*).

Fig. 50. Technique for using long-handled paring chisel.

heads range in width from 3 millimeters (1 *rin*) to 4.84 centimeters (6 *bu*). A medium-sized chisel is 1.515 centimeters (5 *bu*) in width; this is the half-*sun* measure which is a common "rule of thumb" in traditional construction.

Paring Chisels *(tsuki-nomi)* Paring chisels (fig. 49), also called slicks, are used to finish work begun with striking chisels, in particular mortise holes and tenon arms. As their name implies, they are guided by hand pressure, not by striking with a hammer. The end of the wooden handle is, therefore, gently rounded to fit comfortably into the palm of the hand and does not require the steel hoop used to strengthen the handle of the striking chisel. The handles of larger paring chisels may be as long as forty centimeters to provide for a firm grip and good control. Unlike the seated posture adopted for using the striking chisels, the carpenter works with paring chisels standing up, one foot resting on the timber and the chisel carefully centered beneath the head. Greatest control of the tool is obtained by using the left hand to stabilize the end of the handle and by bearing down while the right hand grasps the handle near the shank in an overhand grip (fig. 50). However, to speed up the operation, it is not unusual for some of the paring of rough surfaces to be carried out using a striking chisel without a hammer (fig. 51).

Fig. 51. A common shortcut of using a striking chisel for paring work.

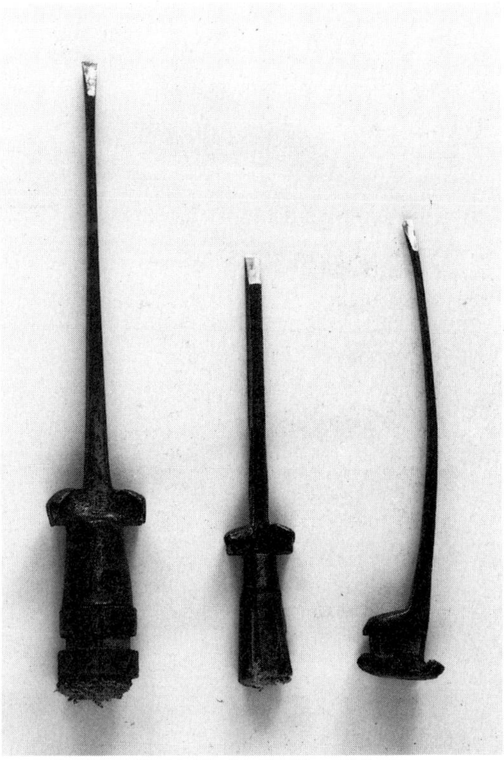

Fig. 52. "Sword-guard" chisels for deep nail holes.

Special-Purpose Chisels Within the broad family of Japanese wood-working chisels there are a number of chisels designed for special purposes including more than eighty different types of gouges used by sculptors. Four types of special chisels are of particular importance to the carpenter.

"Sword-guard" chisel (tsuba-nomi) (fig. 52) Contrary to popular belief, iron nails are used in Japanese building construction. Often square in the shaft and up to twenty to thirty centimeters in length, these nails fasten together many of the vital parts of a building, such as the raftering and eaves framing (fig. 53). The sword-guard chisel is designed specifically to cut the deep, square holes needed for these nails. The blade of the tsuba-nomi is square in section and tapers toward the head, thus creating a hole into which the square Japanese nail is rammed.

It is difficult to extract a standard striking chisel after it has been sunk as deeply as twenty to thirty centimeters into a dense-grained structural timber. The sword-guard chisel is the answer to this problem. By striking it repeatedly on alternate sides in the direction of the handle, the chisel is inched out of its predicament.

Fig. 53. Early nineteenth-century roof cantilevers held in position by deep-sunk nails, as shown during restoration and rebuilding in 1987. Palace of the Inner Citadel, Nijō Castle, Kyoto.

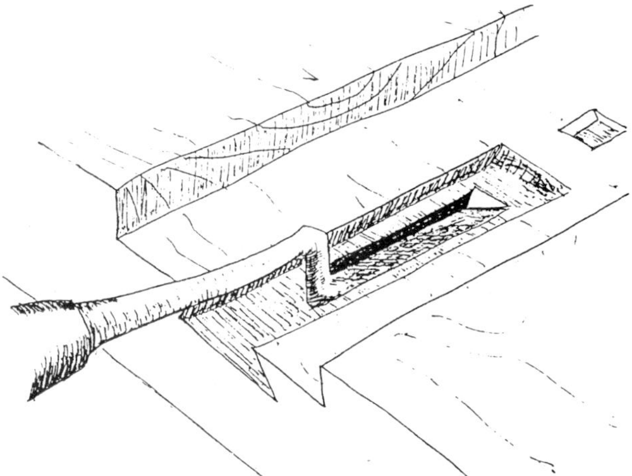

Fig. 54. Trowel chisel (*kote-nomi*) used to fashion a sliding dovetail.

"Bottom-cleaning" chisel (sokosarai-nomi) This hook-headed paring chisel (fig. 42) is used for cleaning waste out of the bottom of a mortise. It is designed to avoid damaging the brittle cutting edge of striking chisels caused by levering them back and forth.

"Strike-through" chisel (uchidashi-nomi) This striking chisel (fig. 42) has a head with no cutting bevel. It is designed simply to break through the wood when cutting a pass-through joint.

"Trowel" chisel (kote-nomi) This paring chisel (fig. 54) has a stepped or chanked shank resembling a trowel handle or a dogleg. It permits the cutting of trench-shaped, sliding dovetails recessed into the bottom of mortises, particularly important for fitting structural braces.

Saws *(nokogiri)*

The next major family of Japanese tools is the saw. Like the members of any family, Japanese nokogiri share common attributes but are also intriguingly different from each other. In fact, the saw has evolved from an inefficient, thick-bladed implement, more effective for its grinding than its cutting action, into a diverse and highly specialized family of precision instruments with varied blade shapes and teeth configurations used for different types of woods and cutting actions. The physical appearance of the Japanese saw and the method of its manipulation are now very different from those of its distant relatives, the European and American saws. The most radical point of divergence is that it cuts on the pull not the push stroke.

Japanese saws generally consist of a wooden handle and a metal blade with teeth down either one or both edges. The single-edged blade is commonly known as *kataba* (literally, "teeth on one side") (fig. 55), while the double-edged saw is called a *ryōba* ("teeth on both sides") (fig. 56). The handle is straight or slightly angled and is made of cypress or cedar bound with bamboo cane. It thus may be used with either a single-handed or double-handed grip, unlike the modern Western counterpart, which is limited by its enclosed handle to a single grip. The blade is made of laminated steel, usually four sheets forged together and carefully tempered and tensioned for flexibility. Better-quality saws are manufactured from high-grade steel. A few saws are still made from superlative "sword" steel using traditional hand-forging techniques. The best quality hand-forged saw with a 2 *shaku* blade (approximately sixty centimeters) reputedly cuts through three hundred logs before resharpening is necessary.

The family of Japanese saws is divided into two main branches: ripsaws (*tatebiki-noko*) and crosscut saws (*yokobiki-noko*). Some roughing saws with coarse teeth straddle both groups, being used for both crosscutting and ripping. The double-edged ryōba, for instance, has ripping teeth down one side of the blade and crosscutting teeth down the other.

The fundamental difference between cutting along and cutting across the grain of a piece of wood makes for important differences in physical form and method of use between crosscut saws and ripsaws. Since ripsaws cut with the

Fig. 55. *Top to bottom*, crosscut saw (*yokobiki-noko*), ripsaw (*maebiki-noko*), small ripsaw used for difficult access, and a heavy-duty ripsaw (*gagari*) with an adjustable handle.

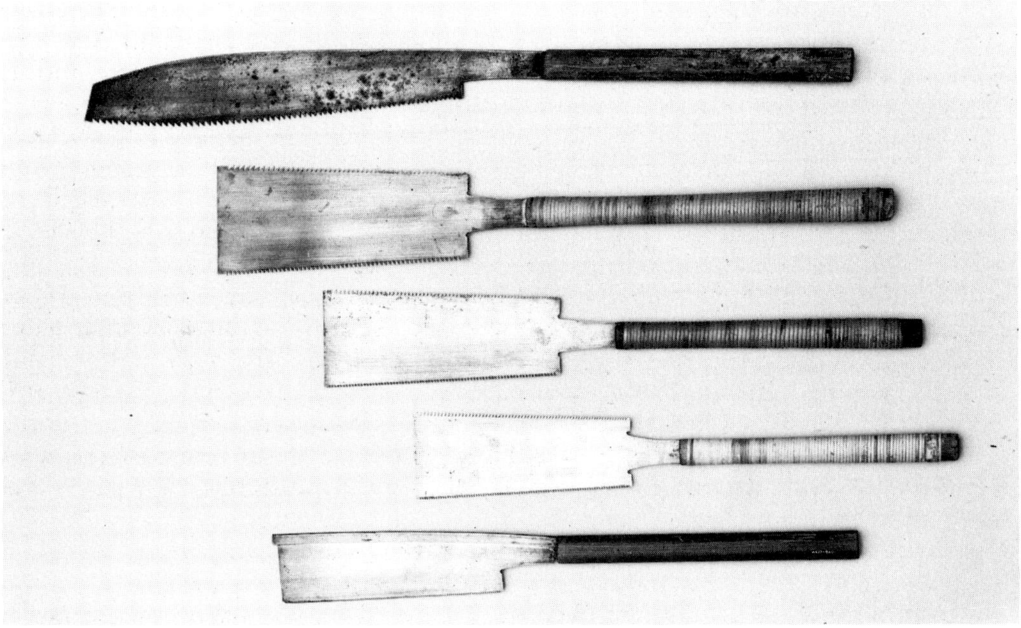

Fig. 56. *Top to bottom*, rounded-nose roughing saw (*hanamaru-noko*), three double-edged utility saws (1 *shaku ryōba*, 9 *sun ryōba*, 8 *sun ryōba*) and tenon saw (*dōtsuki*).

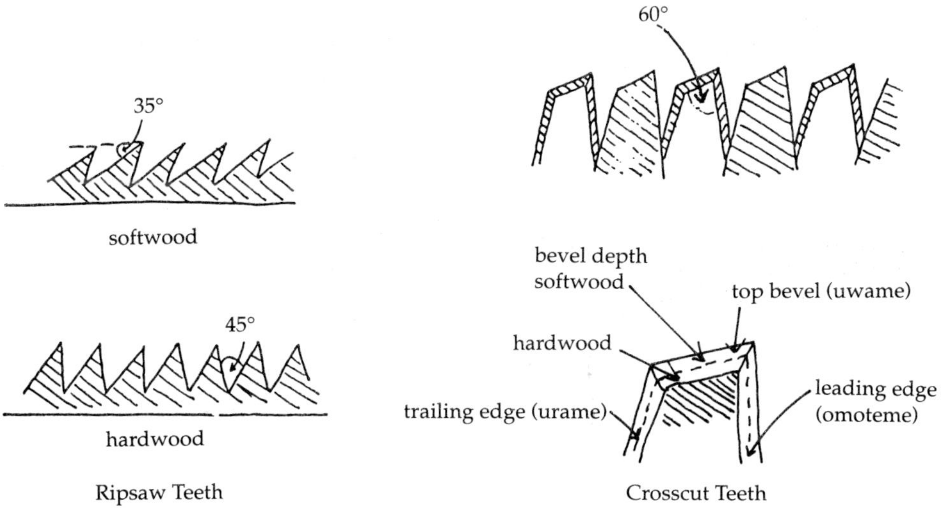

Fig. 57. Teeth configuration for crosscut saws and ripsaws.

grain of the wood, they need large, sharply raked teeth (fig. 57) which look rather like thorns, hence their name "thorn teeth" (*ibara-me*). The danger of teeth breaking increases toward the end of the blade. To counteract this, the angle of the teeth reduces toward the tip. Teeth nearest the handle are set at an angle to the blade of approximately ninety degrees for harder woods and seventy-five degrees for softer woods, but saws used on harder woods have an angle of only forty-five degrees at the toe. The individual teeth of a ripsaw work in a manner similar to the blade of a plane, cutting readily with the grain of the wood at a shallow angle. Unlike the teeth of crosscut saws, they do not require a beveled cutting edge because they work with, rather than against, the direction of the wood fiber. Those teeth nearest the handle, however, are much larger than those close to the tip of the blade and are used at the beginning of each pull stroke to start the cut.

If the teeth of the ripsaw may be said to work like a plane, those of the crosscut saw may be said to work more like a chisel. Its teeth have to chew laboriously through the fiber of the wood at a right angle, unlike the ripsaw teeth, which run freely down the grain. To facilitate this process, it is equipped with smaller, flat-topped teeth beveled on the top as well as the leading and trailing edges for better cutting. In order to make it easier to cut through the wood, the leading edge of each tooth stands practically vertical, compared with the gently raked teeth of the ripsaw.

Japanese saws are thus equipped with highly specialized teeth designed to cope efficiently with different types of cutting actions as well as wood hardness. In fact, the general philosophy for using a saw is to "use the teeth, not your strength." The carpenter's role is to guide these teeth to their requisite position, where they duly carry out their predetermined task, not to exhibit his physical

Fig. 58. Technique for using a crosscut saw.

Fig. 59. Technique for using a ripsaw.

strength by forcing the saw through a resisting piece of wood. To best carry out this task, the saws of the master carpenters are customized cutting implements. They are made by the saw-smith with one particular carpenter in mind. The character of the saw, its liveliness in the hand, is molded to suit the character of the carpenter. The smith spends much time paring down the blade at the center and toward the handle to give it the right "feel."

To use the crosscut saw, the carpenter steadies the wood with his left foot while holding the saw with his right hand. His head is centered over the cutting point of the blade, and the saw is held at an angle of approximately forty-five degrees to the wood. He uses the left hand to guide the blade while the right pulls the blade in short, sharp cutting strokes. When the cutting slot is long enough to stabilize the blade, the left hand holds the base of the handle loosely, and the saw is moved horizontally through most of the length, as gentle power is applied for the pull stroke (fig. 58). The ripsaw, by contrast, is held at a steeper angle of about thirty degrees with feet straddling the wood or right foot resting on it, the left foot pressing against it from the side (fig. 59). The hands are held further apart than for the crosscut saw in order to permit greater control over the blade as it moves rapidly down the grain.

One further general point of fascination to observer and specialist alike needs to be made. Although the same fundamental distinction between ripping and crosscutting is observed in Japanese as in Western saws, the Japanese have chosen a basically different approach to the actual task of sawing timbers. The teeth of the Japanese saw face back toward the handle, indicating that it cuts on the pull stroke. Western saws cut on the push stroke, except for bow and frame saws which can be used on either the push or the pull stroke. Even Chinese and Korean saws, from which the Japanese saw was originally derived, cut on the push stroke, a fact that singles out the Japanese saw as exceptional among those of every known building tradition.

There are general advantages in pulling rather than pushing a saw. Pulling puts the blade of a saw under tension and makes it stronger; pushing a saw places the blade under compression and gives rise to a tendency to buckle and bind when passing through the wood. It is interesting to note that the old Western bow-saw used a blade kept under tension by a tightened cord and could be used with either a push or a pull stroke, depending on the personal preference of the user. Although it has been likened to sawing with a picture frame, the superior cutting performance of its narrower blade still make it the preferred saw for some dedicated craftsmen. A further reason that the Japanese may have adopted the pulling rather than pushing stroke for sawing is that craftsmen tended to work sitting down. Ripsawyers, for example, did much of the arduous lengthways cutting of timbers into long planks sitting down on the job, as may be seen in a photograph taken in 1871 (fig. 60). Here they sit beneath the timber, which is propped up at an angle of about forty-five degrees. To the Western mind, accustomed to the image of a sawyer standing erect over the trunk of a tree, this position seems to offer nothing but aching arms and a cramped back. However, the pull-stroke saw may be operated efficiently and smoothly in a sitting position beneath the timber. Sitting gives a stable fulcrum for the cutting action, ensuring great accuracy down the full length of a beam, while from his position beneath the beam the sawyer gains maximum assistance from the force of gravity in his downward pull-cutting stroke. Many Japanese woodworkers, especially the carpenters who specialized in making furniture, shoji frames and sliding-door frames, generally work seated to free their legs and feet to hold and support the timber being cut.

The special character of the Western saw derives from the logic of the push-cutting stroke, in the same way that the character of the Japanese saw is derived from its pulling action. The blade of the Western saw is thicker than that of the

Fig. 60. Sawyers using a wide-bladed ripsaw in a seated position. *The Far East*, 1871.

Japanese saw and is taper ground so the top edge is thicker for extra strength. The Japanese saw needs only a slight hollowing at the center of the blade to reduce friction in the cutting action. The teeth of Western saws are not as steeply angled as Japanese saws, since greater resistance is met in pushing rather than pulling a saw through timber. The teeth of both Western and Japanese saws are set alternately to each side, but on the Western saw the set is wider, resulting in a wider cut, or kerf, to allow the blade to slide readily. The set of the teeth on a Japanese saw is designed more for discharging sawdust than facilitating the cutting action. The thicker Western blade with wider set teeth and wider kerf requires greater effort in use and results in greater destruction of timber fibers, even though great care is exercised.

A pulled saw is also easier to control than a pushed saw. The hands tend to center automatically as they are retracted toward the body. Consequently, a straighter line is held. It is notable that for this reason the starting strokes in Western usage are themselves made on the pull stroke. Only when the cut has been established is a pushing action adopted. The one disadvantage of the pull stroke as used in Japanese practice is that more sawdust may be deposited on the surface of the wood before it is cut, thus obscuring guide marks. This is not a serious disadvantage, since the pulled saw always cuts straighter.

The steel of the Japanese saw is harder than that used for the typical Western saw. This is not a fortuitous result of differing technology but is the consequence

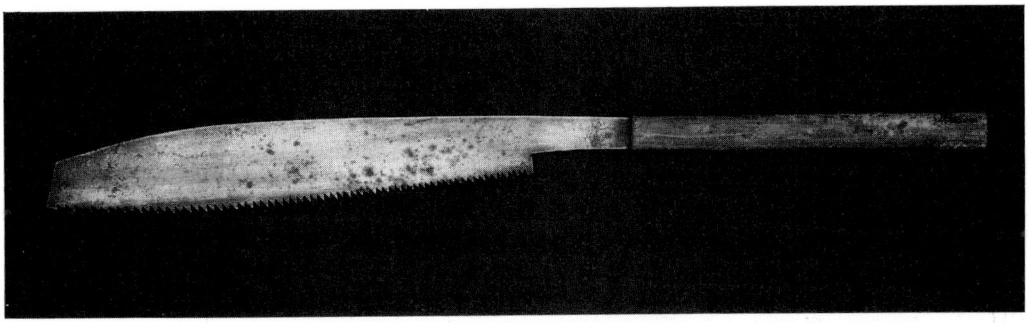

Fig. 61. Rounded-nose roughing saw (*hanamaru-noko*).

of the two different cutting actions. Western saws, on the one hand, are made from softer, more flexible steel to allow for buckling under the stress of a push stroke. Because the steel is softer, the teeth themselves cannot take as sharp an edge as the harder steel of the Japanese saw. The drawback, as it were, of the harder-steel Japanese saw is that it is more brittle and the teeth more liable to snap off when it is used or sharpened inexpertly, or if a crosscut blade strikes a strong knot. However, the saws seem to be very forgiving, as they work just as well with several teeth missing. The major types of saws are set out below.

Rounded-Nose Roughing Saw *(hanamaru-noko* or *anahiki)* This general roughing saw (fig. 61) takes its name from the characteristic droop at the top of the blade. The cutting edge is "breasted," that is, it curves slightly outwards to enhance the cutting action by following the natural curve of the carpenter's arm as it swings in the shoulder socket. High-quality Western saws are also breasted for the same reason. The hanamaru-noko in most frequent use has a blade approximately forty-five centimeters long and 10.5 centimeters wide (1 *shaku*, 5 *sun* by 3 *sun*, 5 *bu* in traditional measurements) and can cut wood ranging from nine to thirty centimeters in thickness. The most common type, the only Japanese saw designed both for crosscutting and ripping, has large teeth, which in design strike a balance between the two teeth types. These teeth are pointed, as in ripsaw teeth, but the leading edges are almost vertical like those of crosscut saws. They increase slightly in size toward the tip of the blade, a modification of the typical ripsaw, but are beveled on the leading edge of each tooth, a modification of crosscut teeth. Some rounded-nose roughing saws have the thorn-ripping teeth only, but these are beveled on the leading edge to permit crosscutting. However, like all compromises, this saw is a less effective instrument in most respects than that intended solely for ripping or crosscutting.

Standard-Purpose Carpentry Ripsaw *(maebiki-noko)* The maebiki-noko (fig. 62) is used in the general course of construction for working small- and medium-sized squared timbers ranging from three centimeters to twenty centimeters in thickness. The blade measures approximately thirty-six to thirty-nine centimeters in length (1 *shaku*, 2 *sun* to 1 *shaku*, 3 *sun*) and is about twelve

Fig. 62. Standard purpose ripsaw (*maebiki-noko* or *tatebiki-noko*).

centimeters (4 *sun*) wide at the tip. It widens toward the top. The back of the blade is noticeably concave near the handle to make it easy for the carpenter to distinguish it at a busy building site from the convex-backed crosscut saw.

The teeth of the maebiki-noko are fashioned in the typical ripping manner, larger at the tip of the blade and diminishing in size toward the handle. The end of the blade is also thinner to reduce friction during the pulling action.

Heavy-duty Ripsaw *(gagari)* The gagari (figs. 55, 63 and 64) is the ripsaw employed principally for use on structural timbers, especially in cutting the tenons of the larger beams. The blade is usually about thirty-six centimeters long (1 *shaku*, 2 *sun*) and twelve centimeters (4 *sun*) wide at the tip. The heavy-duty ripsaw is identical in shape and teeth configuration to the standard carpentry ripsaw, except that it is inserted into the handle at an angle of approximately forty-five degrees. This permits efficient cutting of larger, denser timbers reserved for the structure of the building by positioning the blade at a more effective angle.

Standard Crosscut Saw *(yokobiki-noko)* The yokobiki-noko (fig. 55) is the general-purpose saw for crosscutting. It is thirty-six centimeters (1 *shaku*, 2 *sun*) in length and eleven centimeters (3 *sun*, 7 *bu*) wide at the tip. Thus it is almost identical in size to the heavy-duty ripsaw, yet has standard crosscut pattern teeth that are beveled on all three edges. These teeth are spread evenly along the cutting edge at approximately three teeth per centimeter (nine teeth per *sun*). It is used for cutting timber from three to fifteen centimeters in width, primarily in the preparation phase of construction when the structural members are

Fig. 63. Technique for using a heavy-duty ripsaw (*gagari*).

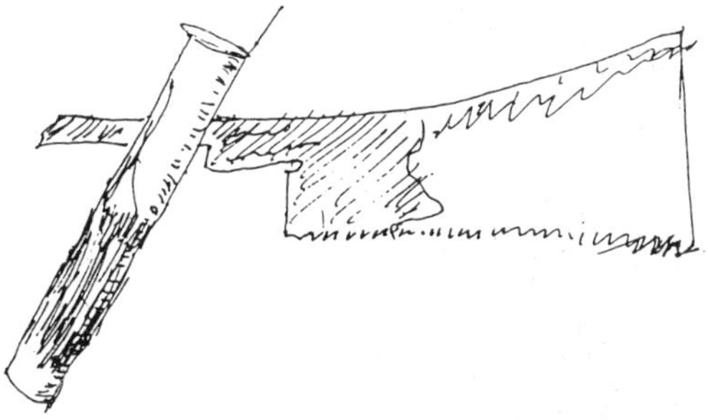

Fig. 64. Heavy-duty ripsaw (*gagari*).

prefabricated. The back of the blade is slightly convex, unlike that of the standard ripsaw, which is gently concave.

Double-edged Utility Saw *(ryōba)* The wedge-shaped utility saw, or ryōba (fig. 56), has cross-cutting teeth down one edge and ripping teeth down the other. Its distinctive shape arises from practical dictates, not aesthetic caprice. The blade widens toward the top to stabilize the large teeth at the end in the cutting slot. The smaller teeth at the narrow part of the blade nearest the handle are used to start the cutting action and do not, as a result, need a wide blade for stability.

The convenience of the ryōba makes it one of the most common of the saws in use today. A daiku customarily has ryōba of three different sizes. The longest of these measures approximately thirty centimeters (1 *shaku*) and is 11.5 centimeters wide with about four teeth per centimeter (twelve to thirteen teeth per *sun*). The long cutting edge of twenty-seven centimeters is most effective when used for cutting timber from three to twelve centimeters in width. The medium-length ryōba, twenty centimeters (9 *sun*) long and ten centimeters wide, has five to six teeth per centimeter (six teeth per *sun*); the smallest saw is about twenty-four centimeters (8 *sun*) long and nine centimeters wide with six to seven teeth per centimeter (twenty teeth per *sun*). The largest saw is used primarily for preparation of structural timbers, and the medium and small saws are used more for delicate finishing work once the structure has been assembled.

Most ryōba blades are slightly concave in section to reduce friction during cutting. Greater thickness at the edges also makes it easier to set the teeth. A few ryōba are actually convex in section. Although harder to use and set, they have a longer working life.

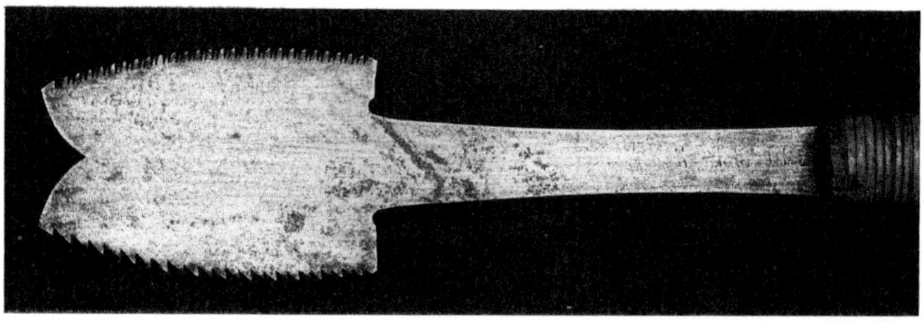

Fig. 65. Intermediate cutting saw (*azehiki-noko*).

Tenon or Back Saw *(dōtsuki-noko)* The tenon saw (fig. 56) is a crosscutting saw ranging in length from twenty to twenty-seven centimeters and usually six centimeters wide. It is designed to cut straight lines with great precision and with a very narrow kerf. The back of the blade is sheathed in a copper spine to brace the blade and ensure the straightness of the cut. This reinforcing permits a minimum set of the teeth, and the kerf may be as narrow as 0.2 millimeters. The exit cut is so smooth that subsequent planing is unnecessary. As the name implies, this saw is used particularly for cutting the shoulders of tenons, but it is also used extensively for detailed finishing work and making shoji frames.

Intermediate Cutting Saw *(azehiki-noko)* This double-edged rip and crosscut saw (fig. 65) is designed to cut midway into a piece of timber. It is especially useful for cutting large curves and sliding dovetails. The blade is short, only about nine centimeters in length, and is set on a long neck. It swells out generously on both edges to let the teeth bite into the smooth surface of the wood. The teeth themselves are very fine, approximately four teeth per centimeter (eleven teeth per *sun*) on the ripping side of the blade and some eight teeth per centimeter (twenty-two teeth per *sun*) on the crosscutting side.

Coping Saw *(hikimawashi-noko)* The coping saw has a long, pointed blade specially suitable for cutting curved patterns into the wooden panels used for transom and other interior decoration. The teeth have no set, which minimizes kerf.

Planes *(kanna)*

The Japanese plane consists of a rectangular block of oak into which is cut an aperture or "mouth" to take the cutting blade (fig. 66). The blade is carbon steel tipped and normally used in conjunction with a capping iron ("chip breaker"). This increases the overall efficiency of the planing action by directing wood shavings away from the cutting edge of the blade. The combination of superlative cutting steel, skillful technique and careful setting of the blade results in long, continuous ribbons of shavings, which may be less than 0.03 millimeters in thickness, and a polished wood surface on which water pools as if on stone. Sandpaper should not be necessary to perfect the surface. In fact, sandpaper is still regarded as anathema by strict Japanese carpenters who believe using it violates the fiber of the wood (fig. 67).

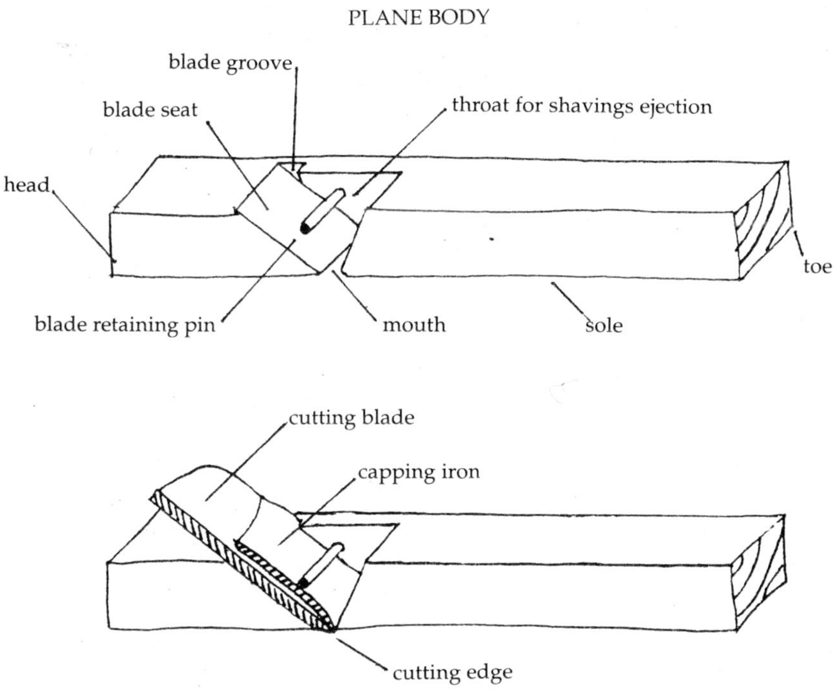

Fig. 66. Plane nomenclature.

Fig. 67. Carpenter using a block plane for finishing a surface.

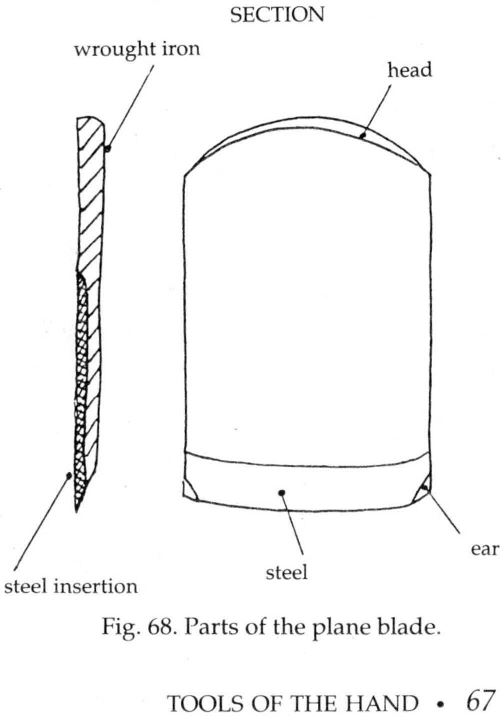

Fig. 68. Parts of the plane blade.

As with all Japanese cutting and finishing tools, the heart of the plane is its cutting iron, consisting of a steel tip forged onto a wrought-iron blank, and honed to razor sharpness (fig. 68). Carpenters frequently made their own special-purpose planes for molding and beveling without the assistance of a blacksmith. In recent years, broken hacksaw blades of Western origin have also found a new lease on life in this role.

A special thin-bladed crosscut saw, the *hoso-noko*, is used to cut the slots at the sides of the mouth of the block that house the cutting blade. This saw is less than thirty-six millimeters in thickness because of the confined space in which it has to work. The teeth are small, with a minimal set designed for cutting the harder wood of the block.

As in the case of the saw, the carpenter works the plane with a pulling action, giving a greater measure of control over his tool than is afforded by the pushing action used in the west and in China and Korea. In traditional practice, the plane was used from either a seated position or standing with the wood set at an angle on a planing board. Carpenters favored the seated position for more detailed work with planes as well as saws. Altogether, it is considered easier to work with a pulling action when seated, a fact which may account for the Japanese plane being used in this distinctive way (pl. 6). Propping up timbers on a planing board supported by a triangular horse made it easier to plane with a smooth, even action, since the force of gravity helped the plane block slide down the timber. However, the angle of the wood was of critical importance: ten to fifteen degrees to the horizontal was considered ideal.

Skillful use of planes is as much a matter of the accurate setting of the blade and selecting the right plane for the task at hand as it is skill in working the tool over the surface of the wood. Japanese planes, like saws, are sensitive instruments which rely upon precision rather than strength to achieve their ends. At the outset, the cutting blade and capping iron must be properly positioned for the particular type of wood and desired cutting strength. Setting the blade is a matter of patient coaxing with a hammer. The alignment of the blade is achieved by sighting along the sole of the plane and tapping the end of the block gently until the required amount of cutting edge protrudes beyond the plane's mouth (fig. 69 and pl. 6). Judging alignment is, of course, a matter of experience. The blade is then locked into position by tapping the capping iron into place.

Similarly the selection of the right plane for the right job entails much experience, especially when confronted by the almost bewildering variety of members of the plane family. The 1943–44 Institute for Labor Science survey list of forty different types of planes as the absolute minimum requirement for a fully equipped master carpenter may seem a surprising number to us today, but we should remember that a similar wide variety of planes was in common use by carpenters, coopers, cabinet and coach makers in Europe and America in the nineteenth century. As Eric Sloane comments on a tool advertisement of the early 1800s, "the army of old-time wooden planes seems overwhelming."[4]

Since the army of Japanese planes is likewise overwhelming (fig. 70), greater ease in understanding may be achieved by dividing them into two broad divisions: smoothing planes for preparing timber surfaces and special purpose

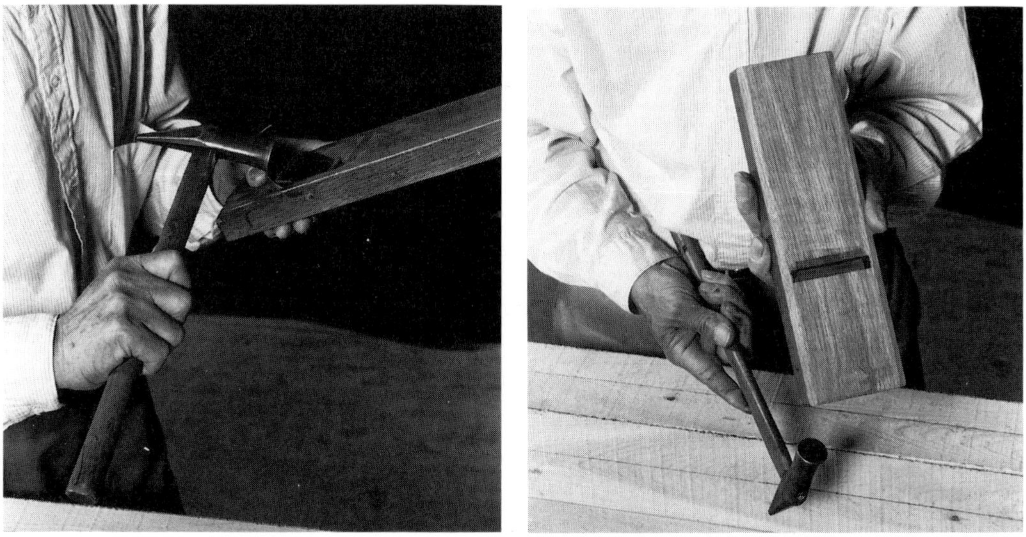

Fig. 69. Setting the blade of the block plane.

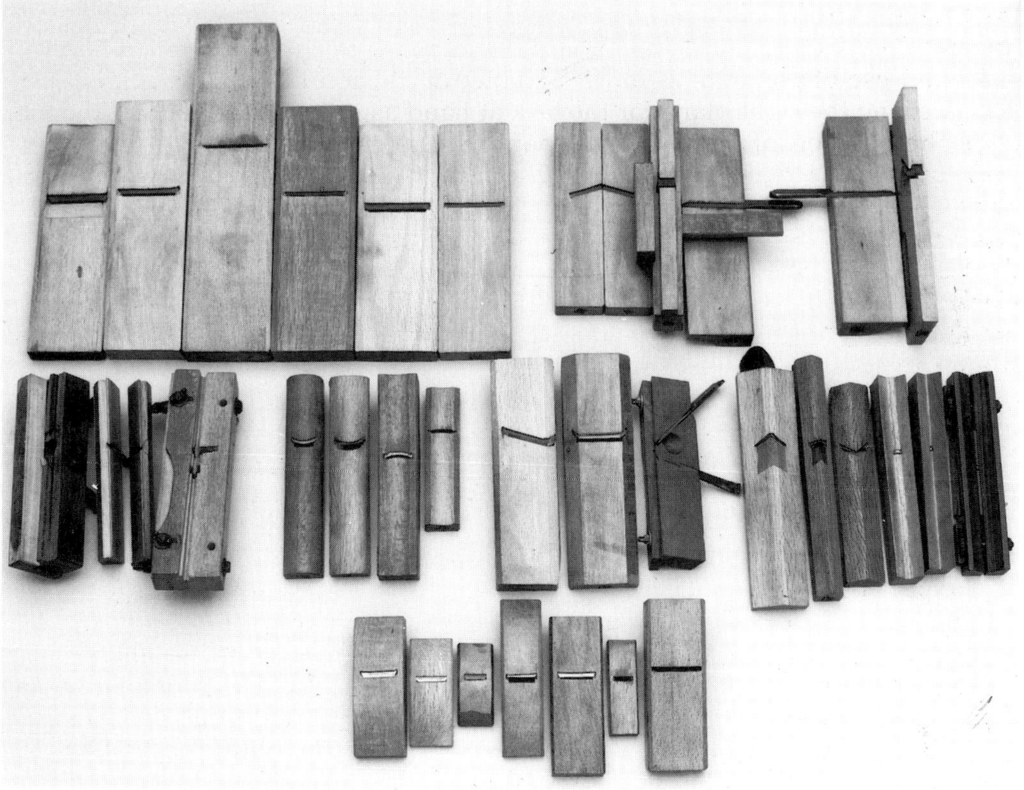

Fig. 70. Some members of the plane family.

Fig. 71. Smoothing planes (*hira-ganna*): *left to right*, preliminary roughing plane (*oni-arishiko-ganna*); general roughing plane (*arashiko-ganna*); jack plane (*nagadai-ganna*); intermediate smoothing plane (*chūshiko-ganna*); finishing plane (*shiage-ganna*); fine finishing plane (*jōshiage-ganna*).

planes for producing the variety of bevels, moldings and track grooves that are used in the traditional interior.

Smoothing Planes *(hira-ganna)* Smoothing planes (figs. 71 and 72) are used to create the seemingly miraculous timber surface effects for which Japanese buildings are renowned. To start the process, a carpenter uses two types of roughing planes to pare down the rough surface of the wood into a more workable form.

Preliminary roughing plane (oni-arashiko-ganna) The major surface irregularities are aggressively cut away with the preliminary roughing plane, the oni-arashiko-ganna. The cutting blade, which needs no capping iron because of the natural fragmentation of the coarse shavings it cuts, is slightly tapered at the cutting edge to increase its actual cutting power. It can remove wood in substantial shavings up to 1.5 millimeters in thickness. The plane's wooden

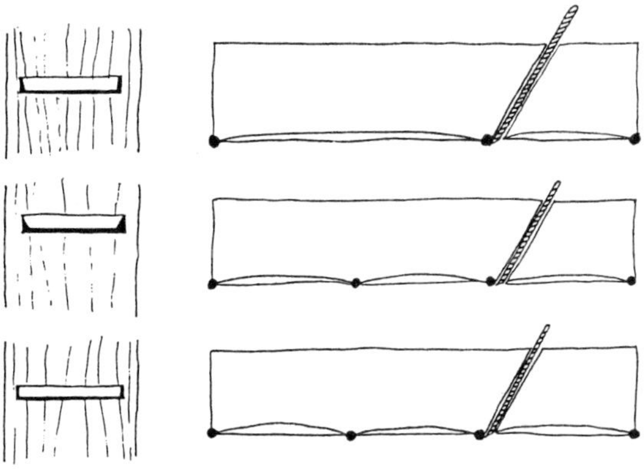

Fig. 72. Comparison of blade cutting edge shapes and plane block sole contact points. *Top to bottom,* roughing plane, intermediate smoothing plane and finishing plane. Concavity is slightly exaggerated for clarity.

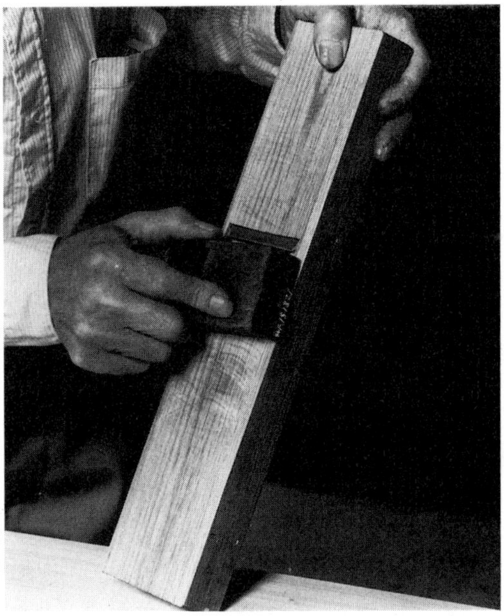

Fig. 73. Scraping the sole of a plane to create concavities for better sliding action.

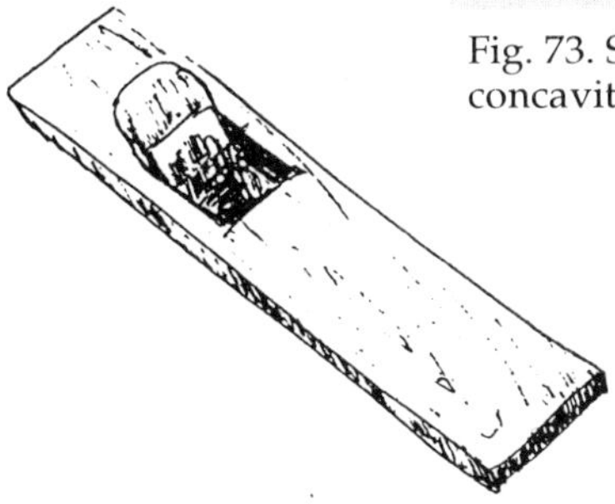

Fig. 74. Jack plane (*nagadai-ganna*).

block is narrower than that of other smoothing planes (4.25 centimeters to 4.85 centimeters; or 1 *sun,* 4 *bu* to 1 *sun,* 6 *bu*) and the sole is planed into gentle concavity on both sides of the blade to reduce drag over the uneven surface of the wood. This tool modification is carried out by the carpenter himself to suit his own planing technique. He uses a small plane with its blade inserted at right angles. This works with a scraping action on the dense, hardwood block to create slight concavities in the sole (fig. 73).

General roughing plane (arashiko-ganna) The second general roughing plane, the arashiko-ganna (fig. 71), is used to complete the rough finishing of the surface. It has a slightly wider and larger body than that of the initial roughing plane and it is equipped with capping iron to ensure it rides with an even, flowing action. This roughing plane is also used at the beginning of the preparatory stage to trim the ends of a piece of timber at right angles to prevent splintering during the longitudinal planing process.

Fig. 75. Grooves for shoji sliding screens. Katsura Imperial Villa, Old Shoin. Early seventeenth century.

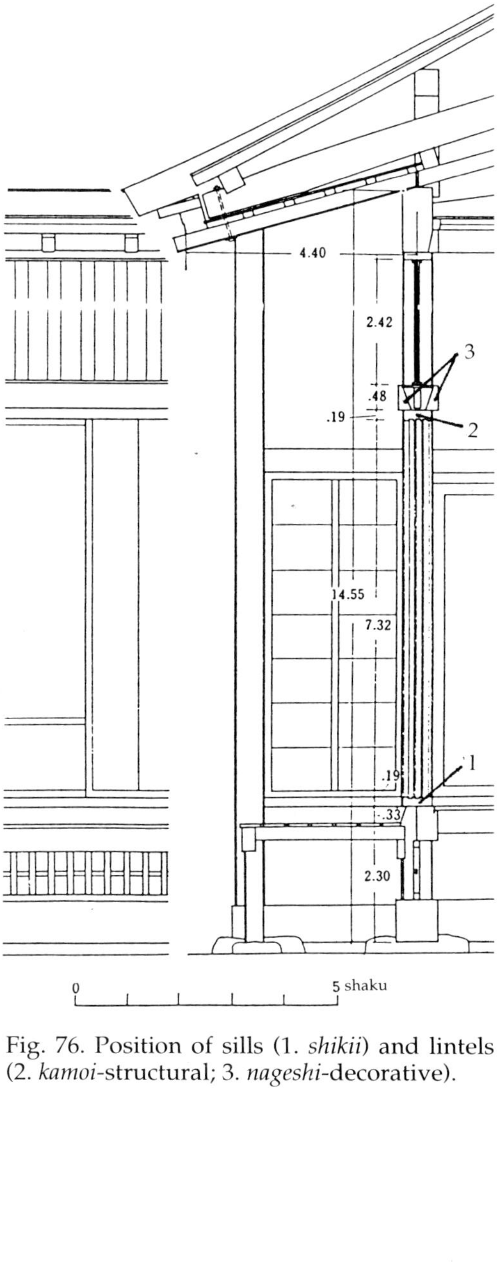

Fig. 76. Position of sills (1. *shikii*) and lintels (2. *kamoi*-structural; 3. *nageshi*-decorative).

Fig. 77. Use of gauges to verify the configuration of the plane sole.

Finishing planes (chūshiko-ganna *and* nagadai-ganna) The timber is now ready for creation of a finely finished surface with the intermediate smoothing plane (chūshiko-ganna) and final smoothing or "jack" plane (nagadai-ganna) (fig. 74). These planes have more sharply tapered cutting blade edges than do the roughing planes, and their soles are designed to allow four points of contact with the wood surface. These differences make provision for more precise cutting and continuous operating stokes down the entire length of the timber. The nagadai-ganna has the longest body of any smoothing plane. A typical length is 45.45 centimeters (1 *shaku*, 5 *sun*), and some planes can be as long sixty centimeters (2 *shaku*). The block of this plane may also be up to 5.43 centimeters (1 *sun*, 8 *bu*) wide. The long, wide block ensures an even final surface on long squared members such as pillars, beams and floorboards.

Special finishing planes (shiage-ganna *and* jōshiage-ganna) The finish on details such as sills and lintels plays an important role in the overall aesthetic effect of a Japanese building. (figs. 75 and 76). Two special finishing planes, the shiage-ganna and jōshiage-ganna, are used to put the final surface on these timbers by bringing out the patterns of the grain. The bodies of the planes are wide (as much as 6.06 centimeters, or 2 *sun*); their blades are not tapered for deep cutting because they do not penetrate the wood surface but rather lightly skim it. The shavings may be as thin as 0.03 millimeters with the lovely translucence of fine shoji paper.

Anyone accustomed to working with wood will readily appreciate that the process of smoothing timber to an even surface is critical both to the final appearance and to the success of much of the joinery in Japanese architecture. Although the flatness and evenness of surface are easily judged by the experienced eye, the carpenter also relies on a special measuring gauge, the *awase-jōgi*, to ensure that the timber is planed flat. These simple wooden straight-edged gauges are also used to verify the configuration of the sole of a plane (fig. 77).

Special-Purpose Planes (fig. 78) The number and variety of special-purpose planes indicates the highly specialized and personalized nature of Japanese carpentry tools. Many of the planes in this category are made by the carpenter himself for fulfilling a special purpose and creating a special effect. An entire book could be filled with descriptions of these gentle and clever tools, but this discussion must be limited to those used most commonly in traditional building practice.

Corner-cutting planes (kiwa-ganna) These planes seem like any other small smoothing planes except that the blades have a raked cutting edge specially designed for creating the sharp corner intersections in joinery (fig. 79). Kiwa-ganna are divided into those raked to cut left-side corners and those raked for right-hand side operation. However, carpenters have been known simply to reverse a plane and use a pull action to cut the opposite corner.

Grooving planes (anazuki-ganna) Anazuki-ganna are used to make the groove

Fig. 78. Special-purpose planes: *left to right*, fenced plough plane (*motoichi-jakuri-ganna*); right and left sill side-shaving plane (*waki-ganna*); "bottom-shaving" plough plane (*sokotori-jakuri-ganna*); right and left corner-cutting planes (*kiwa-ganna*).

Fig. 79. Corner-cutting plane (*kiwa-ganna*) to make sharp joinery interfaces.

tracks, in which sliding screens and shutters run, and the narrow mortises essential for lintel assembly. The Japanese building is equipped with fusuma, thick paper screens which divide rooms, shoji, which form the outer boundary of rooms, and solid wooden exterior shutters (*amado* and *mairado*), which protect the house against rain and wind. All these screens must ride smoothly in their grooves because they are in constant use. Accurate, straight and even tracks must therefore be cut into the wooden top and bottom sills into which they slot. Three types of grooving planes are used to perform this task.

First, the fenced plough plane (*motoichi-jakuri-ganna*) is used to rough out the

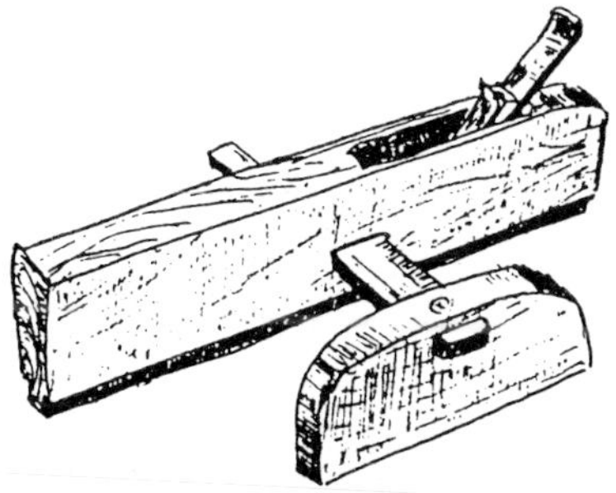

Fig. 80. Fenced plough plane (*motoichi-jakuri-ganna*).

Fig. 81. Technique for using fenced plough plane (*motoichi-jakuri-ganna*).

Fig. 82. "Bottom-shaving" plough plane (*sokotori-jakuri-ganna*).

groove along guidelines laid down with the double-headed marking gauge (fig. 80). It has an adjustable outrigger arm and guide block that sets the plane on an even track parallel to the edge of the timber. The plane has a narrow body with the sole stepped in at the base to 1.515 centimeters (5 *bu*), the standard thickness of a sliding screen track (fig. 81).

Second, the side walls of the track are smoothed to an even finish with side-shaving planes (*waki-ganna*). The angles of the blades located in the side of the block are individually set for left- and right-side cutting (fig. 78).

Third, the bottom of the track is completed with the "bottom-shaving" plough

Fig. 83. Chamfer planes (*kakumentori-ganna*).

Fig. 84. Molding planes (*soridai-ganna*).

Fig. 85. Convex and concave grooving planes known as "rounds" (*sotomaru-ganna*) and "hollows" (*uchimaru-ganna*).

plane, the *sokotori-jakuri-ganna* (fig. 82 and pl. 8). This plane has a very tall and narrow body designed to fit snugly into the 5-*bu* groove to stabilize the plane against sideways wobble. These planes expel shavings from the side because they are too high for a standard top aperture to do the job efficiently.

Chamfer and molding planes (kakumentori-ganna *and* soridai-ganna) The beveled corners of square pillars and lintels, which are a subtle but significant feature of Japanese architecture, require their own special beveling or chamfer planes, some of which are very intricate in design. Bevels or chamfers are more than just an aesthetic feature of Japanese architecture. They are used to eliminate many of the sharp intersecting wooden surfaces and so make the buildings safer for their users and appealing to the eye. The width of the bevel is determined as part of the proportional order of a building. The bevels of pillars of large-scale gateways and residences may be one-tenth the width of the pillar. Those on pillars of smaller buildings and residences may be as narrow as one-fifth of the pillar's width.

There are many different types of chamfer planes, or kakumentori-ganna (fig. 83 and pl. 7). Each is designed to cut a bevel of a certain width and angle. The sole of the block is cut into the shape necessary to guide the blade into the required path. Simpler planes have a sole cut at right angles so that the plane fits snugly to the corner of the pillar to ensure a perfect finish. More complicated planes may have several fluted grooving tracks, and the blade is serrated to cut multiple grooves. Some planes have twin blades arranged like pincers, designed to cut the sliding dovetails indispensable for joining flat pieces of wood.

Another important type of plane are the small soridai-ganna. These come in many variations and are used to make many of the curved edges and rounded surfaces that complement the right-angled post-and-lintel system of the building frame, particularly the curved window frames and ceiling support struts. A molding plane has a rounded sole that cuts into the surface of the wood to shape it into elegantly curved window frames, struts and beams (fig. 84). The contour of the sole itself has a prescriptive effect on the character of the curve.

Most molding planes cut with the grain down the length of the piece of wood, but the convex-grooving plane (*sotomaru-ganna*) and the concave-grooving plane (*uchimaru-ganna*) (fig. 85), known in the West as "rounds" and "hollows," cut across the grain to make rounded tracks in the surface of the wood.

The carpenter makes many of his own chamfer and molding planes. Often they are only finger-size and comprise a single cutting blade without capping iron. They are one of the most subtle expressions of the personal nature of the art and craft of the Japanese master carpenter. Looking at the variety of these planes makes it possible to appreciate the intimate link between human creativity and the means of expressing it in built forms.

Gimlets *(kiri)*

Gimlets are mainly used for cutting holes in bamboo and other interior detailing for sukiya architecture, and for inserting the thread hole into the body of the sumitsubo. The round handle is up to twenty-five centimeters in length, almost

Fig. 86. Gimlets (*kiri*).

the same length as that of the Western gimlet, and is rotated rapidly to left and right between the palms of the hands. The hands thus move under pressure rhythmically down the handle until it is necessary to return to the top and repeat the process. The cutting heads are thin and squared in section with a variety of cutting tips ranging from pointed to barbed in shape (fig. 86).

The Japanese carpenter differed from his Western counterpart in that he never cut large, round holes into the wood for structural purposes. In early times there was some need for round holes in the ends of the trunks of large timbers for transportation purposes. Ropes were inserted through the holes to lash timbers together to make rafts for floating down rivers. However, round holes were unnecessary for structural work. Mortises were completely cut by chisel, not begun with "starter holes" as in Western practice. The holes required for sinking larger nails had to be square because the nails themselves were square in section to prevent them from rotating or slipping under stress. These holes were cut by a sword-guard chisel or with a forked-head square gimlet. This gimlet is known as the *uchikomi-giri* ("striking gimlet"). Since it is square in shape and struck with a hammer, not rotated by hand, it could be categorized as a chisel rather than as a gimlet.

Tools for cutting round holes were imported from the West in the later nineteenth century. The Labor Science Survey reveals that by the 1940s the large-handled gimlet and the standard Western brace and bit had become a regular part of the equipment used for building carpentry.

Hammers

Hammers (fig. 87) vary greatly in size, shape and materials precisely according to function. They differ from their Western counterparts in one significant aspect: they have no claws for extracting nails. This task is performed by the

Fig. 87. The hammer family: *left to right,* two wooden mallets (*kizuchi*); steel double-faced hammers for use with chisel (*gennō*); carpenter's hammers (*kanazuchi*).

nail-extractors (fig. 89). The three main types of Japanese hammers are wooden mallets, metal double-faced hammers and the carpenter's hammer.

Wooden Mallet *(kizuchi)* These are similar to European mallets but range in size from small-scale tapping instruments to large mallets with handles over a meter long (fig. 88). Their primary function is to help assemble the frame elements by exerting sufficient force to drive recalcitrant members together without splitting the joints or bruising the wood. The large mallet, more like a wooden sledgehammer, is used for setting the ridgebeam, main pillars and crossbeams.

Steel Double-Faced Hammer *(gennō)* The gennō is used primarily to strike chisels in the process of creating mortises, although the smallest gennō is sometimes used to drive nails. It has a wooden handle of oak or holly, either straight or slighly curved, which is approximately twenty centimeters long. The iron head is cylindrical with flat striking faces of steel. The head is heavy relative to the area of its striking face. The head of the standard-sized gennō weighs 300 grams and the striking surface is twenty-three millimeters by twenty-six

Fig. 88. Use of large wooden mallet (*kizuchi*) in assembling a timber-frame building.

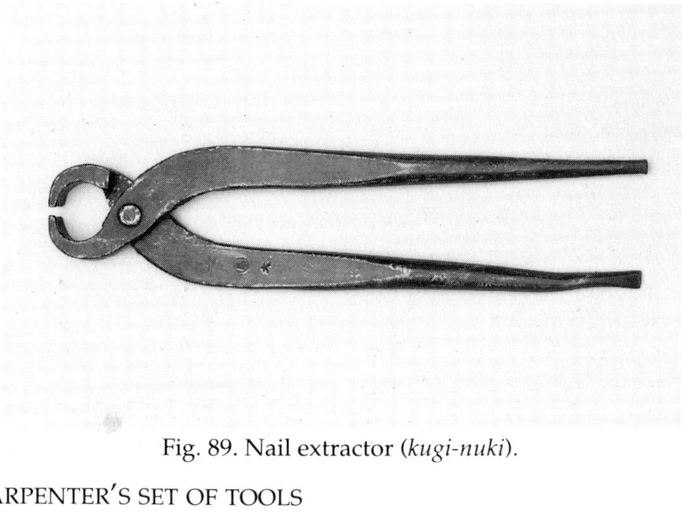

Fig. 89. Nail extractor (*kugi-nuki*).

millimeters, a fact that permits chisels to be driven into the wood with considerable force.

Carpenter's Hammer *(kanazuchi)* The kanazuchi has a metal head with two operating surfaces, one flat and the other tapered sharply like the Western cobbling hammer. This hammer is used for the sole purpose of driving nails, a fact that flies in the face of the commonly held belief in the West that no nails are used in traditional Japanese building practice. As noted before, nails are not used for structural joinery but they do perform a number of functions in the typical building, including securing roof battens and eave end-tiles.

In the nailing process, after the hole for the nail is chiseled, the flat hammer head is used to drive the nail about three-quarters of its length into the wood. The head is then reversed and the operation completed by using the tapered end. In this way, the narrow head of the hammer strikes only the nail and does not bruise the surrounding wood. Thus, the head combines the roles of hammer and nail punch.

Nail Extractors *(kugi-nuki)*

This tool is like a pair of long-handled pincers and, as the name implies, is used for removing nails (fig. 89). It is first depicted in a twelfth-century painting showing scenes of hell, in which miscreants pay for their evil doings by having teeth and tongues efficiently removed by this tool.

SERVICE TOOLS

Some service tools, such as the hoso-noko for cutting the slots at the sides of the plane mouth and gauges for checking the plane sole, have already been discussed. Other service or maintenance tools are essential for the proper functioning of cutting and finishing tools. The most important are the whetstones, used for sharpening the blades of chisels and planes, and the files and rasps needed for sharpening the teeth of saws.

Whetstones *(toishi)*

The razor-sharp edges needed for Japanese tools are created by painstaking and frequent honing. A carpenter may spend many hours each day sharpening his blades. This is essential to realize the full potential of their superlative steel. It is commonly said that four hours of sharpening are necessary to allow six hours of cutting on a typical working day. The patience with which the carpenter applies himself to this seemingly tedious task is, in reality, a form of self-discipline and meditation as he prepares himself as well as his tools for the task ahead.

The whetstones (figs. 90 and 91) are as specialized as the tools they serve. The 1943–44 survey listed a minimum of eight whetstones as essential for carpenters' work. In fact, fifteen to twenty is a more common number to be found at a building site. The stones are usually kept immersed in water ready for use. They

Fig. 90. The daily discipline of blade sharpening.

Fig. 91. Whetstones (*toishi*) *left to right*: from coarse to fine.

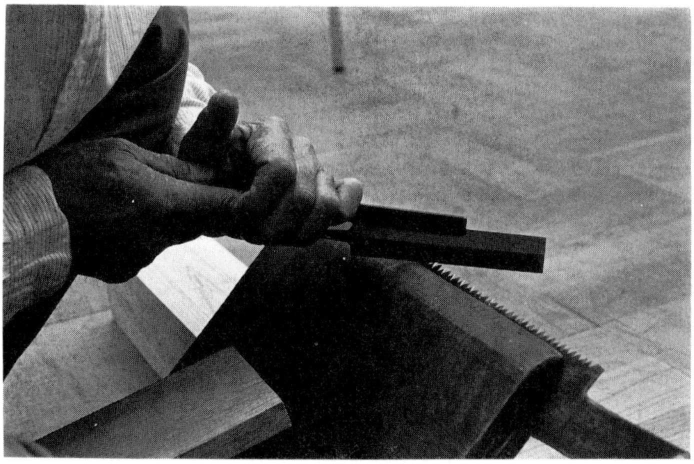

Fig. 92. Filing saw teeth.

are mounted on a wooden base to steady them and are used lengthways for maximum honing effect.

Whetstones range from very coarse (800 grit) to extraordinarily fine (6,000 grit). The coarse stones (*arato*) are used to hone the beveled surface of cutting blades. Medium stones (*nakato*) are the biggest group of whetstones and serve to transform the blades of a tool into fine cutting edges. Finishing stones (*awasedo*) hone the blade to perfection and remove the burr from the back of the blade.

Files and Rasps

The effectiveness of the cutting action of the saw depends upon the care with which its teeth are sharpened. A flat rasp is used to ensure that all the teeth are properly aligned. The teeth are then set to left and right with quick, deft taps of a sharp-nosed hammer. Files which are diamond shaped in section are then used to sharpen each tooth. The teeth of the ripsaw are filed at an angle of ninety degrees to the blade, but those of crosscut saws are filed at an angle of only forty-five degrees (fig. 92). The file dances in a carefully rehearsed sequence from leading edge to trailing edge and back to the leading edge again before filing of the top bevel completes the routine.

During filing, the saw is gripped firmly in a portable wooden vice consisting of two thin boards. These hold the blade but leave the teeth exposed. Today it is more economical and efficient to return high-quality saws to the maker for resharpening. It is, moreover, cheaper to buy a new saw than to resharpen a poor-quality saw.

The file is such an unpretentious tool that it is easy to overlook its importance to the carpenter. It is instructive to see this tool through the eyes of Fukuzawa Yukichi (1835–1901), newspaper magnate and founder of Keio University. One of the most vivid experiences of his youth in the 1850s that he recounts in his autobiography is a description of a file maker at work:

> I knew how to make an ordinary file from a steel bar, after a fashion, but the fine file for sharpening saws was beyond my art. . . [W]hen I first came to Yedo (later Tokyo), I was surprised at the sight of a boy, an apprentice to a blacksmith, making a sawfile. . . .The boy had the file on a piece of leather on an anvil, and was chiseling away at very fine notches as if he never realized there was any wonder in it. I stopped and watched him, thinking what a great city of industry this must be where even a youngster could make a saw-file such as myself had never dreamed of making.[5]

TOOLS CLOSE TO EXTINCTION

There are four important tools which have virtually disappeared from the building sites of modern Japan. They are discussed in greater detail in chapter 4 but take their rightful place at this point to complete an examination of the set of tools customarily used in traditional building practice.

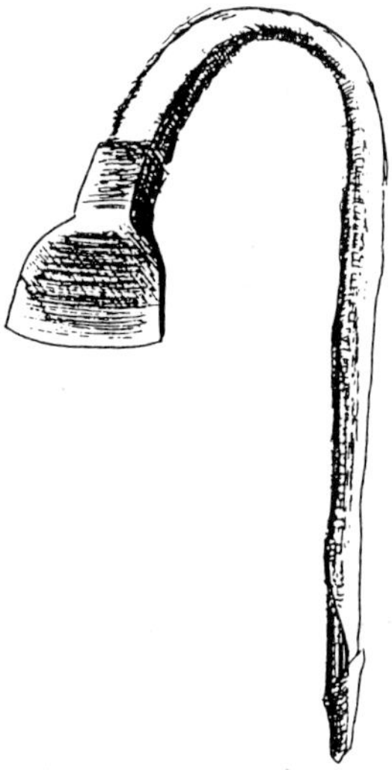

Fig. 93. Ninteenth-century adze (*chōna*). *Morse Collection, Peabody Museum, Salem, Massachusetts.*

Fig. 94. Technique for using the adze.

Fig. 95. A ship carpenter (*funa-daiku*) at work in 1952. Izu-Imaihama. *Photo: Frank W. Coaldrake.*

Adze *(chōna)*

The adze (fig. 93) consists of a curved wooden handle fitted with a heavy, steel–tipped blade usually about ten centimeters wide. The handle was forced into its distinctive curve by repeated heating and gentle pressure to give it the most effective shape for an efficient chopping action. The handle was only about fifty to sixty centimeters in length, making the adze suitable for use seated as well as standing.

The adze was used in premodern building practice for rough finishing the surfaces of pillars and beams and for making hewn planks even. It was particularly useful for fashioning the smooth, rounded surfaces needed for temple and palace pillars. It also performed the essential preparatory work for the fine, silklike surface created by the planing knife or the block plane, although it is common to find the distinctive, dimpled finish of its blade as an attractive feature of castle towers and gatehouses, where smooth surfaces were not essential. The adze takes much skill and patience to use properly and safely. The rough surface of the wood is slowly whittled away with a carefully directed chopping action akin to the pecking motions of a farmyard hen. The wood is smoothed first in one direction and then from the opposite direction (fig. 94). Edward Morse, when he saw Japanese carpenters using adzes in 1877, was amazed at their skill but appalled at the risk:

> It is alarming to see a man hewing a beam standing on it with his naked feet and bringing down his razorlike adze in vigorous strokes within half an inch of his naked toes. Apparently they rarely cut themselves, as I have examined many for the marks of scars, or the gap for missing toes, and in only one case have I found a scar, and when I called the carpenter's attention to it he smiled and showed me another big one on his leg, pointing to his adze and smiling again.[6]

The adze is one of the most venerable of all Japanese carpentry tools, its use dating back to ancient times. It is included in the 1943–44 Labor Science list and could still be seen in general use at building sites and in the wooden ship construction yards of rural Japan, such as at Imaihama in the Izu Peninsula, into the 1950s (fig. 95). The universal adoption of power tools since the end of the 1950s guaranteed its relegation to the museum shelf. The only craftsmen who still use it as a matter of course are the makers of the koto, which is made of paulownia wood hollowed out with a round-headed adze.

Hewing Axe *(masakari)*

The wide-headed hewing axe (fig. 96) was used for smoothing the sides of pillars and beams. Rough hewing was performed with a long, pendulum-like swinging motion, which emitted sounds like the rhythmic ticking of a grandfather clock (fig. 97). Finer finishing was carried out with a two-handed grip on the handle and a carefully directed, gentle chopping action. The adze could also be used as a hewing axe by the simple ploy of removing the head

Fig. 96. Hewing axes (*masakari*).

Fig. 97. Technique for using the hewing axe.

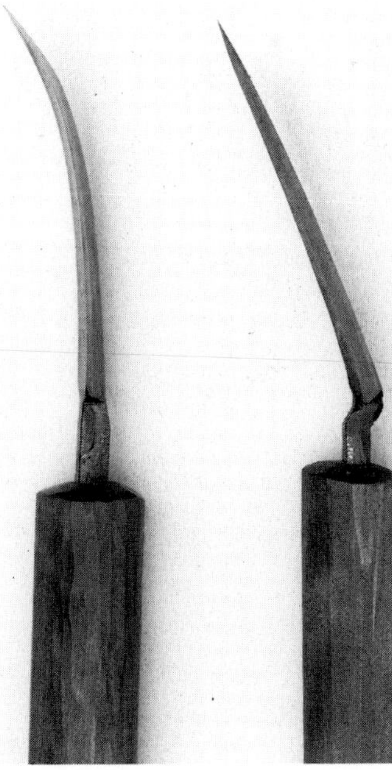

Fig. 98. Planing knives (*yari-ganna*).

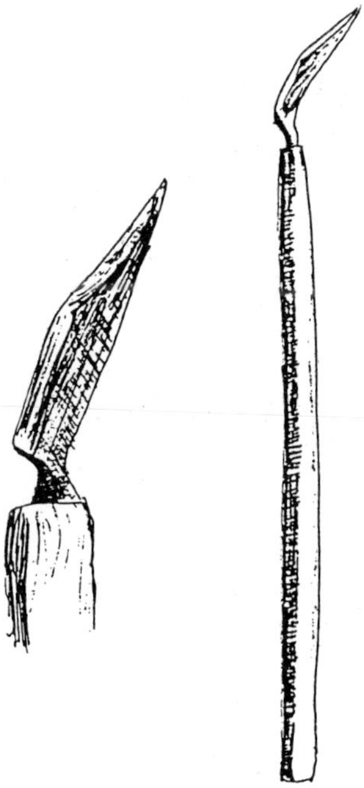

Fig. 99. Angled blades on planing knives (*yari-ganna*).

from its customary position at the tip and replacing it sideways at a ninety-degree angle to the handle.

Planing Knife *(yari-ganna)*

Until the late sixteenth century the yari-ganna was still used in conjunction with the adze to smooth the surface of timbers. The adze was the roughing tool and the yari-ganna, with its long wooden handle and razor-sharp, spear-like blade sharpened on either one or both sides, smoothed surfaces to a silken sheen (fig. 98 and pl. 2).

The yari-ganna was operated normally with long, even strokes, and the cutting edge was angled at approximately thirty degrees to the direction of the grain. The blade was set at an angle of approximately forty-five degrees to the axis of the handle to make it easier to hold, and the cutting edge sometimes had two sections. The main part, from the center toward the rear of the blade, was used for rough finishing; the point area was reserved for fine finishing (figs. 98 and 99). The yari-ganna was used mainly with a pulling action to ensure better control. It demanded the sharpest steel for its cutting edges and the ultimate level of skill from its user.

By the time of the Institute of Labor Science survey in 1943–44, the yari-ganna was no longer part of the carpenter's set of tools. Its use has been revived recently, however, by master carpenters, working at the construction of new

 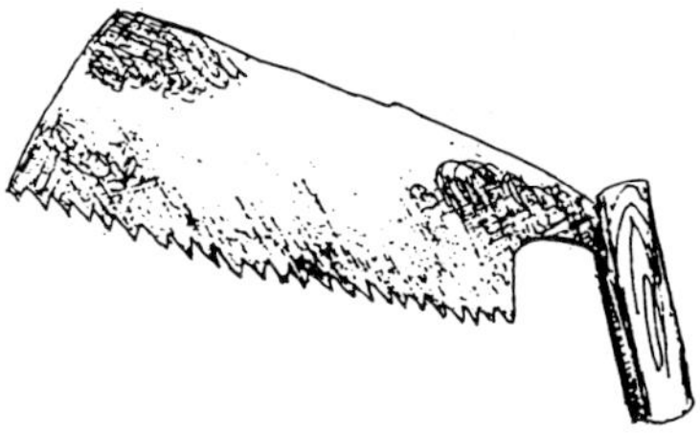

Fig. 100. A ship carpenter checks the teeth of wide-bladed ripsaw (*maebiki-oga*), Izu-Imaihama, 1952. This type of saw is now virtually extinct in building practice. *Photo: Frank W. Coaldrake.*

temple buildings in Nara, who have found that the wood surface it creates is smoother than that made with the finest set of planes. The yari-ganna, although difficult to master, does not share the tendency of the block plane to create gentle hollows in the surface of the wood as it is pressed down to facilitate the cutting action. The surface of timbers finished with a yari-ganna has also proved to be more water resistant than that of timbers finished by plane. Rain actually beads and drips away as though it is running off polished marble. This helps account for the longevity of timbers in parts of the temple of Hōryū-ji near Nara, dating back to the seventh century, which were finished with this tool.

Wide-Bladed Ripsaw *(maebiki-oga)*

This huge, humped-backed saw, often forty centimeters long and twenty-five to thirty centimeters wide, appears idiosyncratic and ungainly, yet it was one of the most important tools in the history of Japanese architecture. It should, therefore, be introduced with the other tools in the carpenter's collection. The maebiki-oga (fig. 100 and pl. 10) was the ripsaw used for preparing large planks and squared timbers, first by carpenters and later by a separate sawyer profession. Only two or three workshops in Kyoto still specialize in its use. As we shall see in chapter 4, its invention caused a revolution in the production and character of the Japanese building.

CHAPTER FOUR

The Invention and Evolution of Tools

This is the "how" and "when" chapter. It looks at how tools evolved in form and when they appeared as an indispensable part of the carpenter's tool set. It also addresses the issue of the way innovations in tool technology and form influenced the design of buildings as well as affected the training and organization of the carpentry profession. These are crucial questions for understanding the evolutionary dynamics of the long architectural tradition of Japan, for the on-site performance of tools influenced the nature of building form as much as any conscious stylistic decisions by the carpenters or the inherent qualities of the wood used. The nature of certain tools also dictated the organization of building activities.

It is important to ask "when" because the Japanese carpentry tools that are familiar to us today represent only the latest stage in a long process of development. The tools which were used to make the temples of ancient Nara were not the same as those used to build the feudal palaces of Kyoto. It would be a serious mistake to imagine that the carpenters who shaped the venerable hinoki pillars of the temple of Hōryū-ji were equipped with today's double-edged saw and versatile block plane. Those seventh-century carpenters would have been puzzled if suddenly presented with modern carpenters' tools because of the great changes that had occurred in the intervening centuries. Indeed, Hōryū-ji would have looked very different had modern tools been used in its construction.

It is essential, therefore, to trace the various stages in the evolution of Japanese tools down through the major historical epochs to the present in order to understand Japanese architecture in the context of the ways and means by which it was produced. To do this it is important to consider tools and architecture in relation to broad changes in society, patterns of religious and political belief, and influences from abroad. The relationship between different tool types needs careful clarification, for improvements in the design of one tool frequently led to modification of others. This approach represents a break with existing studies of tools written in Japanese. These are based on an analysis of

THE EVOLUTION OF JAPANESE CARPENTERS' TOOLS

Period	Buildings	General Trends	Measuring/ Marking Tools	Cutting/Finishing Tools
Yayoi (300 B.C.-A.D. 300)				
		arrival of bronze and iron		chisel, adze, axe, wedge
Tumulus (300-700)	Toro site (Shizuoka Pref.)			gimlet planing knife (*yari-ganna*)
Asuka, Nara (552-794)	Hōryū-ji Tōdai-ji	Sino-Buddhist influence	*sumitsubo;* adoption of Chinese units of measure	crosscut push saw
Heian (794-1185)	Daigo-ji Byōdō-in			
			sashigane with *urame* (square root scale)	leaf-shaped crosscut saw (first push, later pull)
Kamakura (1185-1333)				
		hinoki supplies depleted; *keyaki* substituted		
		first shogunate established; Song influences		
		Ashikaga shogunate	carpenters' manuals	
Muromachi (1333-1573)	Tōfuku-ji Nanzen-ji			two-man frame-saw for ripping (*oga*) (from China)
		earliest extant technical drawings of buildings		
Civil Wars (1467-1573)	Daitoku-ji			

Momoyama (1573-1603)	Azuchi-jō		Japanese invention of one-man ripsaw (*maebiki-oga*)
			single-blade push plane (from China) (*dai-ganna*)
Edo (1603-1868)	Himeji-jō Katsura Imperial Villa Nikkō Tōshō-gū	Tokugawa shogunate	grooving plane pull-action plane
			lumber crosscut saw (*daigiri-oga*); chamfer and molding planes
			Japanese tenon saw utility crosscut / ripsaw (*ryōba*) plane blade with capping iron
Meiji (1868-1912)	Meiji Restoration	standardization of measures	
			Adoption of Western-style nails
	Akasaka Palace Tokyo Station		
Taishō (1912-26)			machine manufacture of saws
Shōwa (1926-89)	Olympic Gymnasiums		domestic production of power tools
			machine production of joinery
Heisei (1989-)			

tools according to type, the evolution of each being traced as a separate species from its origins to the present day. As a result, tools have been isolated from each other and from the formative forces that generated the most crucial developments.

Retracing the evolution of major Japanese tools types is like a detective story. It takes time, patience and a great deal of imagination to piece together evidence from disparate sources to solve the mystery. Few of the many hundreds of thousands of tools used down the centuries in architectural construction are still extant. The special tools used by master carpenters were literally worn out in service. Saws were resharpened until the teeth disappeared, chisel blades honed away to mere stubs. We must rely heavily on pictorial and literary sources, coupled with close examination of extant buildings, some of which date back 1,200 years. The most difficult period for studying tools is that from the eighth to the fifteenth centuries. Archeological discoveries in the burial mounds of rulers prior to this time include some rich tool finds. With the cessation of tomb building in the seventh century, actual tools disappear almost entirely until the fifteenth century. The number of tools surviving from these seven hundred years would barely fill a single carpenter's workbox.

Deductions about tool use can also be made by learning the techniques of master carpenters at work on historic buildings today. This sharpens an appreciation for detail and imparts a certain logic of form and practical process which has evolved over many centuries.

EARLY STONE IMPLEMENTS

Japan during the pre-Jōmon period had its share of stone chopping, scraping and cutting implements dating back to at least 27,000 B.C. From the distribution of these artifacts, it is assumed that man-made shelter was built by Paleolithic man, but firm evidence of building construction dates to the later millennia of the subsequent Jōmon period (ca. 10,000–300 B.C.). This long period witnessed the evolution of a relatively sedentary pattern of life based on hunting and food gathering and a sophisticated ceramic culture. Traces of pit houses dating from ca. 3500 B.C. have been excavated from the large mounds of discarded shells which accumulated at settlements along the coastline. Clusters of pit houses grouped around a common area were also found in the central Japan Alps at the sites of Togariishi, Yosukeone and Idojiri. Judging from the evidence of rebuilding, these villages were in continuous use for several centuries.

A closer look at these pit houses reveals a great deal about the people who built them and the tools they used. The structures were roughly circular or oval in plan and four to five meters in diameter. They had sunken earth floors covered by thatched roof superstructures supported by four to six posts set into the ground. The posts were surprisingly large, usually about forty-five centimeters in diameter. Stone tools excavated at these and other sites show that trees were felled and roughly shaped with axes and adzes which had flaked or polished stone heads (fig. 101). Some of the larger trees were split laterally to make squared members, which are easier to assemble than rounded ones. The

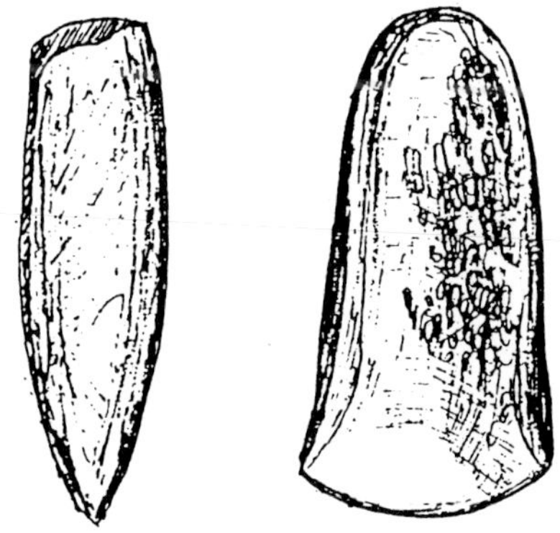

Fig. 101. Polished stone axe or adze heads with breasted cutting edges.

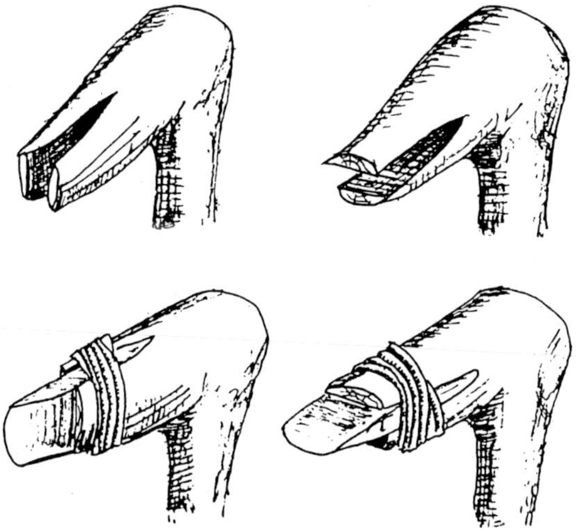

Fig. 102. A stone-headed axe and adze from the Jōmon and early Yayoi periods.

heads of the axes and adzes were lashed to wooden handles (fig. 102). The adze handle was fashioned from a forked branch with one arm shortened for the head and the other serving as the handle. These chopping tools were not specially designed for building work and would have been turned to the tasks of hunting and food preparation as readily as the felling and of hewing timbers.

Jōmon man became highly skilled with his stone-headed axes and adzes, vigorously attacking the huge cedars and other trees found in abundance in the ancient forests. Many of the extant axe and adze heads are large, measuring as much as thirty centimeters in length, seven to ten centimeters wide at the cutting edge and five centimeters thick. They were also designed for maximum efficiency, with breasted cutting edges to enhance the cutting action, and their mass would have given them considerable striking force. Structures of some size were able to be built with these simple tools. Over eighty large pillar stumps dating to 2500–2000 B.C, and still bearing the marks of strenuous efforts with stone axes, have been excavated from the Chikamori site in Shiga Prefecture while others have been found at the sixth century B.C Mawaki site in Fukui Prefecture. At Fudōdō, in Toyama Prefecture on the Sea of Japan, the earth floor and pillar marks of a large rectangular building have been discovered. It measures seventeen meters long and eight meters wide and has sixteen large pillar holes. Four stone-lined hearths are aligned along the central axis of the building. It must have served an important ritual role in the community, judging by the central location.

Stone continued to be used for tools until about the third century A.D., when it was eclipsed by the arrival of bronze and iron from the Asian continent. It is interesting to note that some stone axe and adze heads were imported even before the arrival of the first metal implements from the continent. Large stones ground into smooth planes with cutting edges sharper than those previously existing in Japan and dating to 100 B.C.–A.D. 100 have been excavated from the Ikegami site in Osaka Prefecture. These appear to have been created on the Asian continent with the assistance of iron tools and imported to Japan where their handles were made locally.

THE METAL REVOLUTION AND NEW VERNACULAR BUILDINGS

The use of iron tools as part of sustained building practice in Japan dates to the short but significant Yayoi period (300 B.C.–A.D. 300), when the basis was laid for much of later Japanese civilization in social, political, religious and cultural terms. From the third century B.C. there were major movements of population into the Japanese archipelago from the Asian mainland. These immigrants brought with them the techniques of wet rice cultivation and established stable, agricultural communities with increasingly specialized use of iron tools. Bronze and iron arrived almost simultaneously in the Japanese archipelago. The softer bronze was reserved for ritual implements like mirrors, ceremonial swords and bells (*dōtaku*), while iron was used for more practical agricultural and hard-working building tools. By A.D. 250 the use of stone implements had virtually ceased. Iron tools excavated from sites dating from this period include axes and adzes, chisels and gouges, cutting blades and the long-bladed yari-ganna or planing knives.

The Yayoi settlements of Toro and Yamaki in Shizuoka Prefecture, which thrived in the fourth and fifth centuries, are two of the most important sources of information available about Yayoi period building techniques and tools. In a society based on wet rice cultivation, there was a close relationship between implements needed for agriculture and tools used for building purposes. Extensive systems of embankments and water channels were created for the inundation of the rice fields, requiring axes, adzes, knives, chisels and gouges to make wooden supports for retaining walls. These same tools fashioned wooden farming implements such as spades, hoes and rakes, as well as dug-out canoes for crossing streams and rivers.

The widespread diffusion of these iron tools, especially the chisels and adzes, resulted in dramatic improvements to the pit dwelling (figs. 103 and 104). Excavations at Toro and Yamaki revealed general use of sturdier, squared posts for roof support. Before the introduction of iron tools, posts were roughly hewn trunks, sometimes split but rarely squared. The absence of joinery on Jōmon timbers indicates that they were simply lashed firmly together with straw rope. At the Yamaki site, however, a number of pillars and beams connected with sturdy mortise-and-tenon joints were discovered. It would have been impossible to fashion this type of joinery with stone-headed tools. This joinery technique, prompted by the availability of iron tools, led to a type of housing that was a significant advance on the Jōmon dwelling in size and sturdiness. It also led to a change in living patterns. Most cooking had been done outside the Jōmon house, but it was safe to cook inside the larger, better built Yayoi residence. The

Fig. 103. Reconstruction of Yayoi period pit dwelling at the Toro Site, Shizuoka Prefecture.

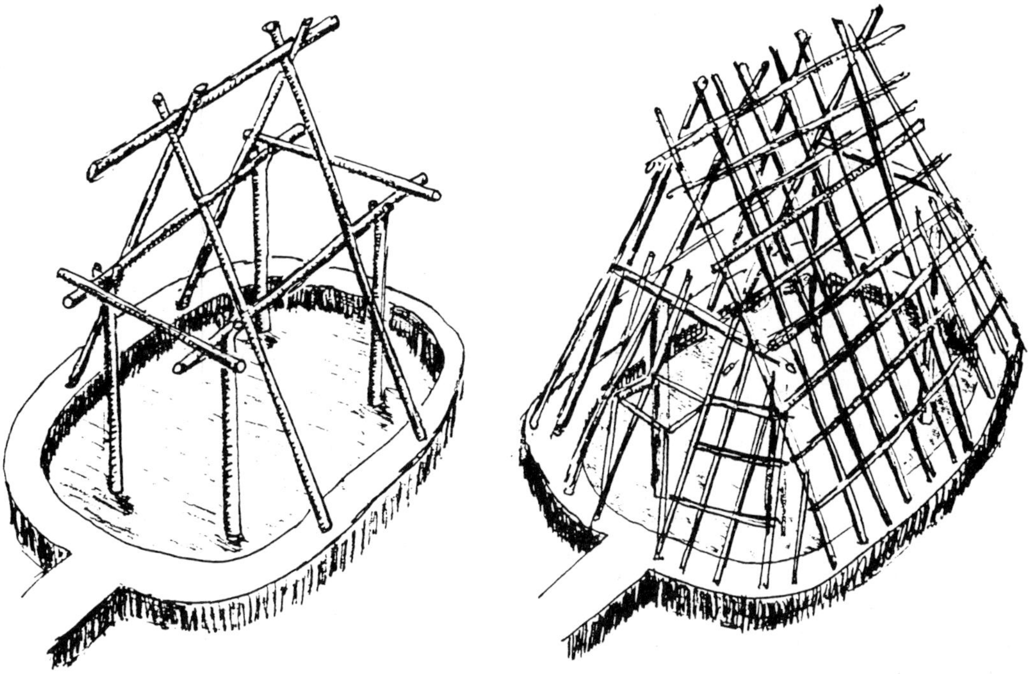

Fig. 104. Structure of pit dwelling.

construction of interior stone hearths became common practice in ordinary dwellings. The durability of the joined structure also transformed construction work from a recurring seasonal or annual activity into a more occasional and professional project. Rethatching with fresh straw or river reeds was still necessary every few seasons, but the frame of a dwelling would last thirty years, an entire Yayoi lifetime.

Although advances in pit dwelling construction were important, the most significant architectural development of the Yayoi period was the advent of a completely new building type, the raised-floor structure (figs. 105 and 106). At the outset it was used for storing the harvested rice safely away from the damp earth of the low-lying areas where rice cultivation of necessity took place, but it later evolved into the residential style for members of the ruling elite and the form for Shinto architecture.

The appearance of these granaries and residences may be reconstructed from archaeological evidence and drawings of buildings incised on a bronze mirror excavated at Takarazuka, near Osaka. The typical storehouse had six posts from three to four meters in length inserted into the ground to the depth of at least fifty centimeters. The walls were made from tree trunks split into planks and held together with log-cabin style interlaced corners (fig. 105). The entrance was customarily set into the side of the building to allow for easy access in storing or removing the grain. The roof was thatched, and its supporting frame was made from light timbers lashed together to create a triangular truss, a distinctive feature being the way in which the rafters projected beyond the ridgepole to which they were lashed. Although the roof differed little in structure and appearance from that of the Jōmon pit dwellings, the frame beneath is clearly the product of efficient iron cutting tools. The posts are longer and straighter than those in use before the introduction of iron tools and much narrower than the great, roughly hewn tree trunks found in the larger scale Jōmon structures. The hewing of the wall planks required axes to split the trees as well as axes and adzes capable of hewing them to an acceptably smooth finish. The interlocking corners and lap joints for the floor joists demanded tools with sharper cutting edges than those available to Jōmon man. Ladders were carved from single pieces of wood and had concave risers hollowed with adzes to prevent rodents running up them. Ladder tops were guarded by baffle boards to turn back rodents that had bypassed the curved risers. These oval-shaped guards required skilled carving to create the necessary center aperture.

The raised floor structures used for residential purposes were similar in most respects to the granaries. They were, however, larger in scale and their doorways were placed in the narrow gable end wall of the building in order to minimize access and maximize protection and privacy.

The raised floor structure was not a Japanese invention. Although recent archaeological discoveries point to the possibility of some form of raised floor structures existing in the Jōmon period, the typical elevated structure of Yayoi times was introduced to Japan from China along with the techniques of rice cultivation. This may be verified by the fact that among the ritual objects found in the excavations of the Shizhaishan tomb, in Yunnan Province in China, built at the end of the second century B.C., there is a cast bronze model of a building

Fig. 105. "Log cabin" (*azekura*) wall construction. Tōdai-ji, Nara Prefecture.

Fig. 106. Raised floor storehouse.

Fig. 107. *Haniwa* house. Fourth–fifth century. Tobi, Sakurai City, Nara Prefecture. *Tokyo National Museum.*

which could be mistaken for a Yayoi-period Japanese granary. The model was probably used as a family shrine dedicated to ancestor worship. It confirms that buildings with the same architectural features as those of the raised floor buildings of Yayoi Japan were used for residential purposes in China at an earlier date. The Chinese model, which was made with considerable attention to detail, shows a sturdy structure with thick, round pillars supporting a raised floor. Oxen are housed in the space beneath the floor while people inhabit the upper level. The walls are made of smoothed planks and have a rectangular opening set into the side to serve as a door. The house is dominated by a sweeping gabled roof which has a series of rafters crisscrossing along the ridgepole, creating the forked effect to be seen in the drawings of Yayoi buildings on bronze vessels. This style of raised-floor building was to become an integral part of Japanese architectural tradition, paving the way for the creation of palatial and religious architecture in the fifth century.

THE EMERGENCE OF ELITE ARCHITECTURE

The establishment of a stable agricultural society in the Tumulus period freed people from the tyranny of subsistence and permitted role diversification, trade specialization and the accumulation of wealth and power. By the fourth century A.D. Japanese society was becoming stratified along occupational and status lines based on extended-family clans (*uji*). One such uji was the Yamato, based in the Asuka region to the south of present-day Nara. It became a powerful clan that achieved a far-reaching hegemony over central Japan through judicious manipulation of political and marriage alliances and military conquest. Yamato leaders were to create the Japanese imperial house, their authority sanctified through association with the mythical Sun Goddess Amaterasu-ōmikami. The Yamato clan sponsored construction of monumental tombs for its dead leaders and increasingly sophisticated residential and religious buildings. The keyhole-shaped tombs, scattered like confetti over what is now called the Kansai region and imitated by leaders in northern Kyushu and even as far east as the Kanto Plain (the modern Tokyo region), have proven to be repositories of immense historical importance. They also disclose important information about building techniques. They consisted of impressive artificial hills faced with stone and containing a number of burial chambers, although only one leader was actually entombed.

The tomb of Emperor Nintoku is the largest and most impressive of these great funeral monuments known to us today. Completed ca. A.D. 427, it was originally surrounded by two moats to separate the sacred resting place of this deceased ruler from the profane world. Although the tumulus is only about 35 meters at its highest point, its base dimensions of some 486 meters by 305 meters exceed those of the largest of the Egyptian pyramids. The proliferation of such tombs has given the period from the fourth to the seventh centuries the name Kofun ("Old Tumulus") period. They indicate that the Yamato rulers were able to mobilize vast armies of craftsmen and workers to create great monuments of state. Although the stone which once covered the outer surfaces of many of

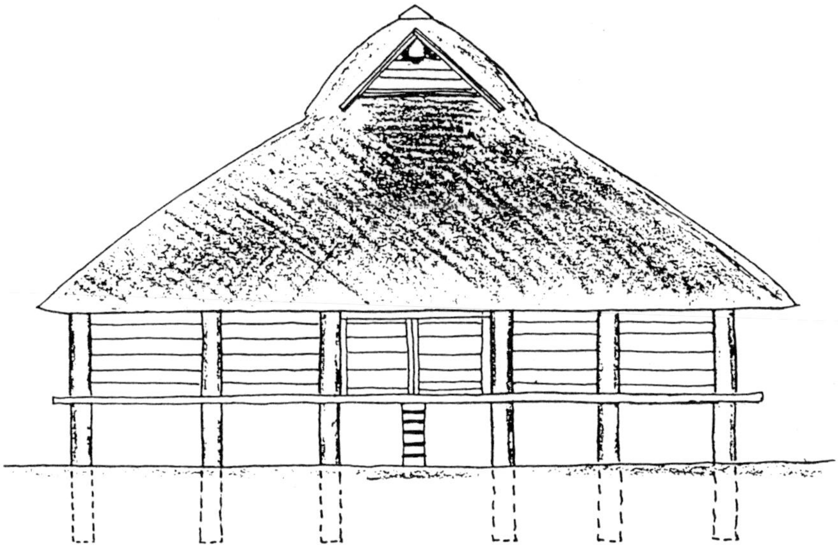

Fig. 108. Reconstruction drawing of a fifth-century aristocratic storehouse (probably Naniwa no Miya) near Osaka Castle.

these grave mounds has long been stripped away, massive dry-wall stonework survives in many of the burial chambers within. This shows that the masons of the day had effective iron-bladed chisels and wedges for quarrying and finishing the surfaces of the rock. The outside of these funeral hills was decorated with hollow earthenware sculpture (*haniwa*) in the form of members of the elite dressed in armor or mounted on horses. There are also ordinary musicians, birds and animals as well as models of raised-floor houses with their distinctive gable roofs and plank walls penetrated by doors and windows (fig. 107).

Archaeological excavations have also revealed that, at the same time as tomb construction was at its zenith, buildings of some size and complexity were constructed at the centers of administrative and political power. This indicates the evolution of increasingly effective woodworking tools. Until the early 1980s, little evidence had been uncovered to indicate the existence of the wooden buildings built by the elite in this period, but the situation is now changing rapidly. Excavations begun in 1986 uncovered one of the most important finds of the century. A group of twelve buildings was found near Osaka Castle at a site immediately adjacent to the assumed location of Naniwa no Miya, the palace of the imperial court in the fifth century. Continuing work has revealed that the buildings, with their floors raised at least one meter above the ground, were probably used as storehouses by the court (fig. 108). All twelve buildings were aligned east to west and oriented on a north-south axis. Each was five bays square, measured approximately nine to ten meters north-south, and was over ten meters east-west. The buildings consisted of two sets of round pillars, forming an inner structure three-by-five bays in size, flanked on each side by a corridor one bay wide. The pillars on the inside were as small as twenty centimeters in diameter, but the outer pillars were twice as thick.

Buildings of such size and consistent dimensions would have required specialized tools and the skills of many craftsmen. Fortunately for us, the nature of these tools is no longer a matter of conjecture. Excavated tombs of the same period have yielded up a number of carpentry tools from their burial chambers, establishing as fact what the remnants of court buildings suggest—that a wide range of tools came into use under the patronage of the Yamato clan. The Kanakurayama tumulus in Okayama City, excavated in 1953 and roughly contemporary with the Naniwa palace storehouses discovered twenty-five years later, contained the most important collection of tools. A total of 48 chisels, 42 gimlets, 25 planing knives, 9 small crosscut saws, 9 adzes, 2 axes and 8 cutting knives were found in four clay pots in the main burial chamber. Their presence in the tomb demonstrates that these tools had ritual significance, perhaps because this technology conferred on the governing elite the power to demonstrate its authority with built monuments.

The Kanakurayama tomb discoveries help to unravel the mystery of what types of tools were used in the building projects of this far-off era. Their number and variations make it clear that chisels were already one of the most important tools in carpentry practice. As to their form, the handles of these chisels were inserted into an iron envelope at the end of the neck of the blade, in a fashion similar to those known to have been in use on the Asian mainland at the same time. A few of them have a tang for inserting into the handle in the style of later Japanese chisels but lack the securing ring that became standard. Many of the chisel blades resemble the hiro-nomi, the more recent wide-bladed striking chisels; the major difference is that the ancient tools have double-edged blades with a pronounced bevel on one side and a slighter bevel on the other. A few single-edged chisels have been recovered from fifth-century sites but the double-edged form appears to have been the norm. The use of a double-edge establishes that these chisels were employed for splitting timbers and roughing out shapes, not for fashioning sharp-edged joints, which subsequently became the primary role of this type of chisel.

Gimlets were the second most numerous of the tools found at the Kanakurayama tomb. These either have straight heads for punching holes or screw spirals for drilling. Then follow the saws, which are small with crosscutting teeth and could have been used only for making small, delicate household objects and personal effects such as combs. The presence of a large number of planing knives is particularly significant. Such finishing tools would have seen extensive service in smoothing the planks used for floors and walls, especially of residential buildings .

The Advent of Shinto Shrine Architecture

A separate but closely linked architectural development of the Tumulus period was the adaptation of the raised-floor structure in the creation of Shinto shrine

Fig. 109. Main Sanctuary, Inner Shrine (Naikū), of Ise Grand Shrine.

architecture. The two major traditions in Shinto architecture of ancient times, the Grand Shrine of Ise (fig. 109) in Mie Prefecture and the Great Shrine of Izumo on the Sea of Japan coast, appear to be based, respectively, on the granary and the residence models. Ise served as the official shrine of the imperial family, housing the bronze mirror, one of the sacred regalia of ancient emperors; the shrine at Izumo remained a center of strong local faith, its political and religious authority eclipsed after the fourth century by the Yamato hegemony.

It is easy to understand how the grain storehouse, which preserved the staff of life, was transmuted into the sanctuary for the gods, the kami, who presided over the universe and its natural processes. Since these granaries were the focus of village festivals celebrating the rice harvest, there was a close physical association with the performance of Shinto rituals. Similarly, the raised-floor residence with its elitist associations was seen as a suitable dwelling place for the gods. The haniwa clay sculptures, which were set into the outside surfaces of the burial mounds, included models of houses which were probably intended as temporary abodes for the kami as they visited the sacred tombs, or for the souls of the departed as they launched on their last journey.

The sanctification of secular buildings placed special demands on carpentry techniques. Both the Ise and Izumo shrines have been rebuilt many times since their foundation, reconstructed as part of a process of ritual renewal. The form of such buildings as they stand today is highly refined and symbolic, and many of the changes occurring during their evolution toward the more abstract were accomplished only with the benefit of the more advanced tools and techniques developed in later eras. However, it must be emphasized that even from the Tumulus period, the separation of sacred architecture from the secular domain demanded increasing refinements in tool use and the quality of wood surface and finish. At Ise the emphasis on refinement led to abstraction of detail, while at Izumo the emphasis was more upon scale. Judging from the size of the stumps of ancient cedars excavated near Izumo, the pillars of the shrine were over a meter in diameter even in early times. Structures of such size would have entailed great strength and skill in construction techniques. It is logical to deduce that with each shrine rebuilding, dictated as much by structural deterioration as Shinto belief, tools assumed increasing religious significance. The importance of smooth, pristine surfaces for ritual purity accounts for the presence of the planing knife among the funerary objects of the fifth century.

THE CONSOLIDATION OF ASIAN MAINLAND TOOLS IN JAPAN

In the middle of the sixth century, monks, scholars and temple builders from the Korean kingdom of Paekche brought to Japan the advanced building styles associated with the Chinese and Buddhist institutions of the Asian mainland. In the Asuka and Nara periods, Japan entered upon an era of conscious modernization using the model of Chinese civilization. Over the next two hundred years, a series of capital cities were laid out in orderly geometric grid patterns in the Asuka-Nara region, and impressive halls and palaces housed institutions of centralized state control based on those of Tang-dynasty China. Buddhism became the official state religion and great monasteries and temples were founded to spread the teachings of the faith of the Buddha throughout the provinces.

Today the grand temple complexes constructed in the seventh and eighth centuries are still the most dramatic evidence of the staggering progress of this era, from the pit dwellings, log-walled houses and raised-floor granaries of a few generations earlier to monuments of unsurpassed creative genius. New styles of building on an unprecedented scale placed new demands on the building professions and their tools. More sophisticated techniques for working with wood, stone, plaster and clay were imperative. Huge Buddhist temples and halls of state called for the ability to work with sturdy framing timbers. The bracket sets that supported the projecting eaves consisted of multiple bearing arms and blocks carefully shaped and precisely fitted together.

The means by which this architectural transformation of Japanese government and society was effected has received little attention. Even the best preserved buildings from the period have been much repaired and rebuilt. Few tools survive and carpenters' documents consist mostly of registers of payments contained in temple archives and some lively graffiti hidden behind the beams of temple ceilings. However, it is clear from early writings such as the *Nihon shoki* (*The chronicles of Japan*), compiled in the first decades of the eighth century, that the earliest temples were built by master artisans from the Korean peninsula who came to Japan for building projects sponsored by powerful court families such as the Soga. Early efforts by the Japanese to emulate continental building styles were tentative: end bays were narrower to compensate for the greater load at the roof corners, and additional pillars were used to prop up eave corners in case the new-fangled cantilevers should prove inadequate. The Japanese already had at their disposal axes, adzes, chisels, gimlets, mallets and planing knives—all the basic cutting and finishing tools of continental practice—from the Yayoi and Tumulus periods. In fact, there may have been fewer immediate changes in building techniques and tool types than is suggested by the dramatic appearance of temples and palaces in the region around Nara. The recent archaeological discoveries of sophisticated palace storehouses and a wide variety of tool types in fifth-century tombs raises the distinct possibility that the techniques which made possible the architectural achievements of the seventh and eighth centuries were not a major departure in construction techniques from the previous centuries. Rather, they were the culmination of a long and logical process.

The best evidence for carpentry tools and techniques for this period comes from Hōryū-ji, the oldest extant Buddhist temple complex in East Asia, and from the Heijō Palace and Tōdai-ji building projects of the city of Nara. The construction of each of these had its own special character and furnishes important insights into the development of tools.

Hōryū-ji—The Temple of the Exalted Law

Founded in A.D. 607 by Shōtoku Taishi (A.D. 573–621), the prince regent responsible for consolidating the institutional power of the Yamato state, the austere beauty of Hōryū-ji established a new standard of architectural taste. It was destroyed by fire in civil conflict of the mid-seventh century but was wholly rebuilt by 710. The West Precinct of the temple encompasses both the famous five-story pagoda and Golden Hall (Kondō), two of the oldest surviving structures, dating to the later seventh century.

The Kondō is an archetypal Buddhist timber-frame building (figs. 110 and 111) comprising a four-by-five bay structure mounted on a stone podium similar to the sanctuaries of Greek antiquity. It is dominated visually and structurally by a massive hip-gable roof and its wide eaves are supported by cantilever bracket sets, which are the most ingenious feature of the East Asian building tradition.

The five-story pagoda, which stands to the side of the Kondō, rises thirty-four meters to the tip of its bronze finial. The pagoda is stabilized against earthquake shock by a great central pillar but the essential structure, like that of the Kondō, is post-and-lintel framing.

The sophisticated overall effect presented by these two architectural masterpieces was created using simple tools little changed from their predecessors a century earlier. It is evident upon close examination of the pillars of Hōryū-ji that carpenters had mastered the techniques of quartering large cypress trunks with wedges. An experienced eye was necessary to judge the character of the grain and the way in which it would split. There are great advantages in this method over even modern techniques for milling timbers. Splitting timbers with the grain makes them resistant to oil loss and rids them of

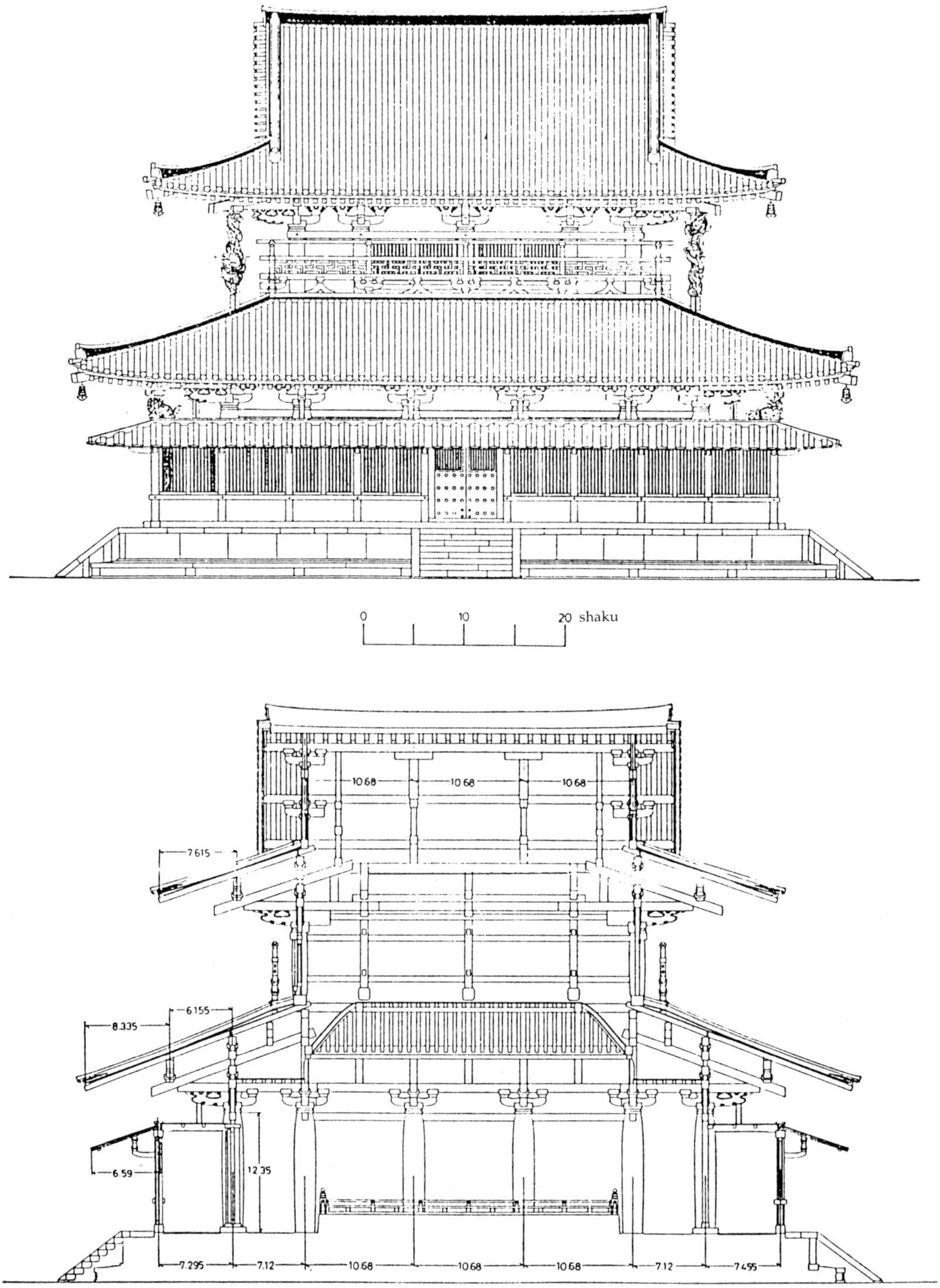

Fig. 110. Front elevation and longitudinal section of the Main Hall (Kondō), Hōryū-ji, Nara Prefecture. *Hōryū-ji Kondō Restoration Report.*

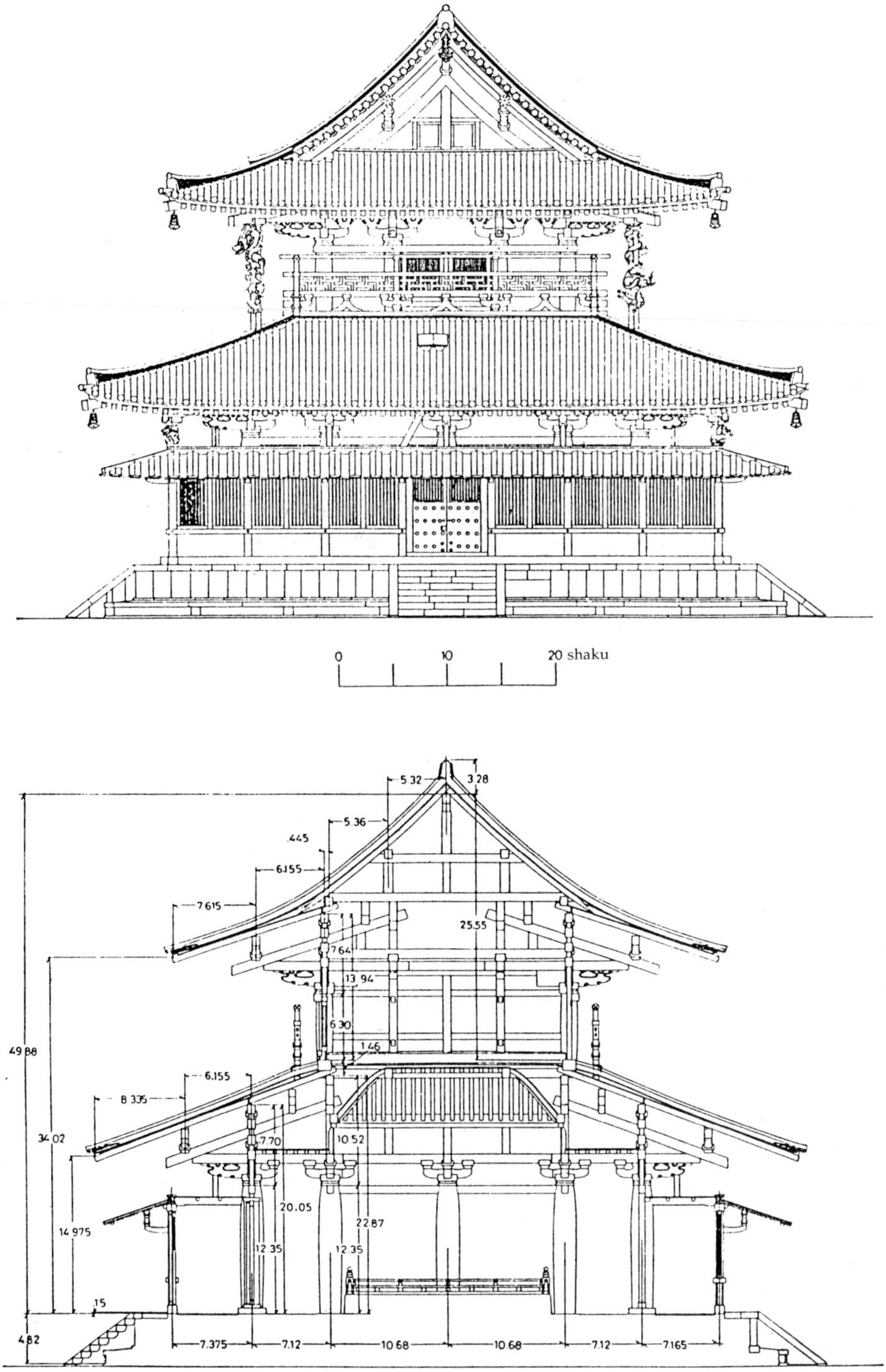

Fig. 111. Side elevation and transverse section of the seventeenth-century Main Hall (Kondō), Hōryū-ji, Nara Prefecture. *Hōryū-ji Kondō Restoration Report*.

Fig. 112. Technique for making round pillar from quartered log: *right to left*, marking into eight planes; rough hewing with an axe (*masakari*); smoothing with an adze (*chōna*); fine finishing with a planing knife (*yari-ganna*).

the propensity to warp, both critical factors in the longevity of the structure.

The quartered timber was subsequently hewed into eight planes, rounded with adze and smoothed to silken smoothness with planing knives to make the round pillars needed for temple construction (fig. 112). Although the pillars fashioned in this manner were carefully shaped and finished, they were sturdy in their proportions, a direct reflection of the tools and techniques used in their creation. It is interesting to note in passing that this technique was by no means unique to Japan. It was standard procedure, for example, to use wedge, adze and drawknife on straight-grained oak when preparing ground sills and other major beams in house-framing construction in Europe. The technique was found still to work effectively in the recent restoration of an eighteenth-century Welsh farmhouse.[1]

Tool marks left on the surface of extant timbers are virtually the only basis on which a picture of tools and techniques used in the construction of Hōryū-ji may be formed. The woodworking tools were chisels, adzes, planing knives and saws. There is also evidence of the use of inking lines and pens as well as carpenters' squares (fig. 113). Although the types of tools used in the construction of Hōryū-ji were limited in number, there appears to have been a great variety in size in each category. The greatest number of tool marks left on surviving timbers was made by chisels (fig. 114); twelve different sizes of blade heads have been identified from marks uncovered on members examined during the restoration project begun in 1934. The smallest of these is 1.8 centimeters wide and the largest is 4.5 centimeters. All the joinery seems to have been cut with chisels. The mortises were cut most frequently with a 2.6-centimeter-wide blade head. Use of the wider-headed chisels appears to have been mostly confined to angled joinery. Curved surfaces and door pin sockets were carved with gouges. Surfaces generally were smoothed with adzes and finished with planing knives, which were also used for finer joinery interfaces (fig. 115).

The adzes used to smooth the surface of the timbers had three sizes of blade head: 9.7 centimeters, 5.5 centimeters and 4.4 centimeters. The largest of these, the 9.7 centimeter blade, became the standard size for all-purpose adzes on building sites until a generation ago (fig. 115). The narrowest blade was used in conjunction with the chisels in joinery work, probably in the one-handed chopping action immortalized in medieval picture scrolls. The marks left by the adze heads in the wood surfaces at Hōryū-ji shows that they struck the timber at right angles to the direction of the grain and at a shallow slant, in contrast with later practice.

Rudimentary crosscut saws were also used in the construction of Hōryū-ji. Small versions of the crosscut saw, excavated from burial mounds of the Tumulus period, were used only for the fashioning of small, delicate utensils and ornaments. Hōryū-ji provides the earliest evidence for the use of crosscut saws in building practice, although they were of limited usefulness. Saw marks are observable only on the butt ends of smaller framing members, such as the inner ends of the cloud-shaped bracket arms and bearing blocks of bracket sets. Some tenons may have been rough cut by saw before fine finishing with chisel and adze. The ineffectiveness of these saws is demonstrated by the existence of several half-sawn timbers seemingly abandoned midway by frustrated carpenters. The kerf of the saws ranges from 0.3 centimeters to 0.8 centimeters, suggesting a grinding rather than a cutting action.

Only part of one tool actually used in the construction of Hōryū-ji survives today. Undeservedly, it is a section of the singularly inefficient crosscut saw type, estimated originally to have measured fifty centimeters in length. It is now preserved in the Hōryū-ji Treasury of the Tokyo National Museum (fig. 116). Experiments have shown that it was extremely laborious in use and heavy in the

◁ Fig. 113. Rear side of the eave-end tile-support beam (*urako*) made of cypress (*hinoki*). This member was cleaved along the grain. The seventh-century sumi markings are still clearly visible on the rear side where they were hidden from view and protected from the elements. They showed the carpenter where the rafter ends and other elements of the eaves structure were to connect with this beam.

◁ Fig. 114. Roof purlin (crossbeam to brace rafters) of cypress with mortise. The width of the chisels used to cut this mortise is still clear today, and the quality of the work is surprisingly rough and ready.

◁ Fig. 115. Large wedge of cypress used to strengthen the joint of a major pillar and beam. The marks left by the adze on the hidden undersurface of the wedge are still clearly visible.

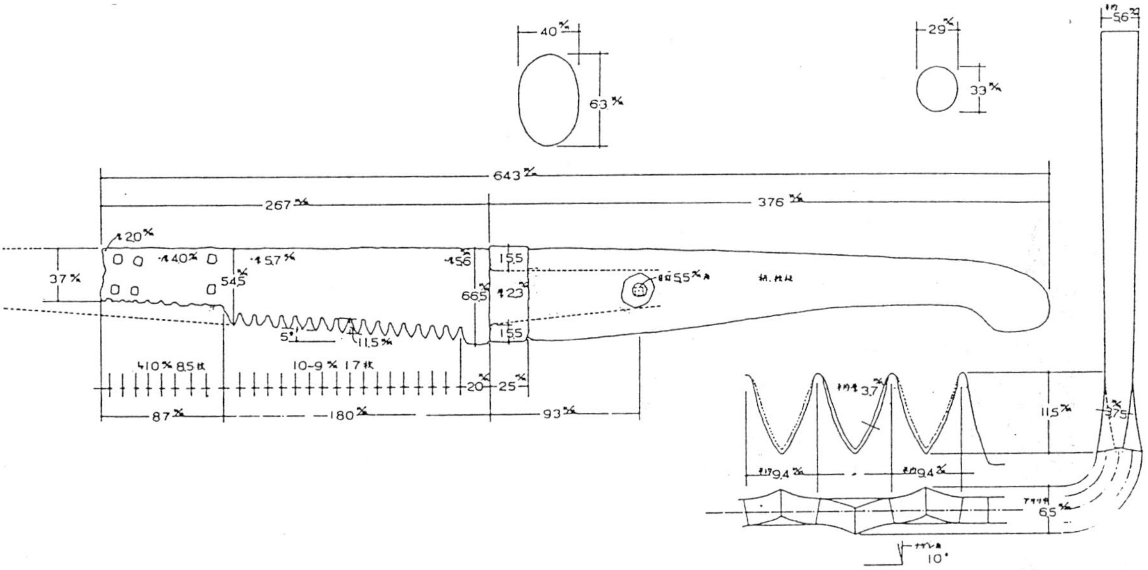

Fig. 116. Crosscut saw from Hōryū-ji. Seventh century. *Tokyo National Museum.*

hand. It was also most effective when used on the push stroke, in conformity with Chinese mainland usage from which this type of saw was derived.

Monuments of the Eighth-century Imperial State—Nara Palace and Tōdai-ji

The construction of the national capital at Nara (Heijō-kyō) in the eighth century symbolized the establishment of a centralized imperial state on the model of Tang China. Laid out in an orderly grid plan which accommodated a population recently estimated at approximately 100,000 people, it was an international center of culture, commerce and religion as well as of government. Its creation as an impressive urban environment was a major national priority and closely

Fig. 117. Reconstruction drawing of the Heijō Palace main palace compound with the Great Hall of State (Daigoku-den) at the center.

supervised at the highest levels of state. During the reign of Emperor Shōmu (r. 724–49), a government memorandum was issued which noted that the capital still had "shingled roofs and thatched dwellings, relics of the building modes of antiquity" and requested that an edict be issued ordering that all courtiers of substantial rank and "all commoners able to do so should build their houses with tiled roofs and paint them red and white."[2] Certainly at the official level the Japanese succeeded in creating major monuments to church and state, even if archeological evidence suggests that domestic architecture fell short of the ideal. The year A.D. 728 saw the establishment of a government department, the Mokuryō ("Construction Bureau"), to coordinate the construction of temples and public buildings and to coordinate urban planning. A special ministry was also set up to build the Heijō Palace, a compound over one kilometer square containing the administrative and ceremonial halls of government as well as the residence of the emperor (fig. 117). In mid-century the building of Tōdai-ji as the headquarters of State Buddhism was deemed so important that another ministry was established to carry out this task (fig. 118). Tōdai-ji and the Heijō Palace were the two most extensive and impressive projects of the Nara period. They were carried out with the same commitment and organization of manpower necessary for staging modern olympic games.

Today these monuments to state and religious authority are but a shadow of

Fig. 118. Model of the Tōdai-ji main compound, eighth century.

their former selves. The Heijō Palace disappeared under paddy fields after the transfer of the capital functions to Naniwa and subsequently Heian-kyō (Kyoto) at the end of the eighth century. Tōdai-ji has lost most of its original buildings to natural disaster or civil war. However, smaller temples like Tōshōdai-ji have survived to indicate what the great palaces and temples of the period looked like at their zenith. Their sturdy bracket sets, elegantly curved beams and gracefully proportioned colonnades represented the height of fashion in the contemporary Chinese cities of Chang-an and Luoyang.

This cosmopolitanism was a singular achievement in so short a time, but what was the technological infrastructure upon which its great monuments were built? Although no tools of the Nara period are known to survive, much can be gleaned about them from the character of the extant buildings. There was considerable scope for carving wood with chisels and planing knives, but the ability to alter substantially timbers of large dimensions was limited by the efficiency and cutting edges of the available tools. Proportions of structural members were massive, and penetration of pillars for insertion of bracing beams was kept to a minimum because anything more would have been too difficult or laborious. Pillars had a simple housing for horizontal bracing cut into the top and a small mortised tie at the base. Heavy squared beams known as *nageshi* were set inside and outside the pillars at base, waist level and lintel height (fig. 119). By the end of the eighth century, these nageshi had became thicker in section, thus allowing the pillars they braced to be made thinner and more elegant to the eye.

Excavations carried out over the last thirty years at the site of the Heijō Palace furnish some useful clues about building techniques. Many pieces of timber framing have been recovered in good condition, preserved for twelve hundred years by the chemical composition of the damp earth of the rice paddies. There is evidence of extensive use of gimlets, especially for making holes for the ropes

Fig. 119. Structural lintels (*nageshi*) of Daigo-ji Five-Story Pagoda, Kyoto. Ninth century.

used to lash timbers together to float them down the rivers from mountain forests. It is clear, too, that the marking and labeling system called *bansuke* was used in eighth-century Japan as part of a system of prefabrication because many of the timbers recovered from the paddies of the past were labeled with letters and numbers as a guide to assembly sequence.

Pillars at the Heijō Palace were smaller in general than those used for the construction of Hōryū-ji but they were made by the same technique of splitting tree trunks into quarters with oak or iron wedges or with heavy duty chisels. The quartered pieces were then rough-hewn by axe. A clue to the next stage of the process is provided by snap line markings which are still visible on some timbers. These indicate that the rough-hewn timbers were pared down with adzes to a carefully predetermined octagonal shape before being rounded off by planing knives into the cylindrical pillar. The pillars of the most important ceremonial buildings, such as the Great Hall of State (Daigoku-den) at the center of the palace complex, rested on individual base stones set on an elevated podium in the Chinese manner. However, the pillars of most buildings, including the residence of the emperor and his family, were set directly into the ground and had elevated floors, in keeping with earlier Japanese practice. The main difference seems to be that pillars of the larger buildings were braced below ground level by penetrating horizontal ties to give additional support for the weight above (fig. 120). This procedure required mortise holes to be cut through the bottom of pillars with chisels.

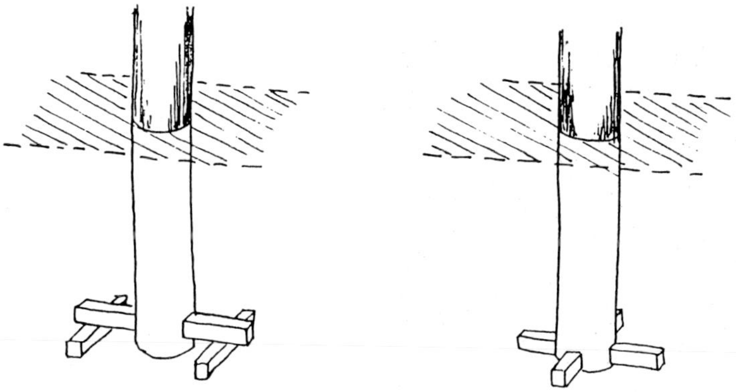

Fig. 120. Lateral bracing of Heijō Palace pillars below ground level.

The crosscut saw was still useful only for cutting timbers less than thirty centimeters thick, less than the diameter of most pillars used in the palace complex. Its most important function would have been for roughing out mortise and tenons and in making furniture, particularly Chinese-style chairs and storage chests that had become popular—even status symbols—for the aristocratic class.

The massive structure of the Great Buddha Hall (Daibutsu-den) of Tōdai-ji today indicates the scale of construction of which the Japanese were capable in the eighth century. A pair of nine-story pagodas, each exceeding 100 meters in height, guarded the approaches to the inner compound containing the Great Buddha Hall (fig. 118). This structure, a colossal eighty-six meters wide and nearly fifty meters high, housed the giant gilt-bronze statue of the Vairocana Buddha, which was the spiritual focus of the temple. Success in erecting buildings on the scale of the Great Buddha Hall and the twin pagodas depended upon the availability of timbers of enormous size capable of yielding readily to splitting, and on efficient organization of human resources, rather than upon any dramatic improvements in tool technology.

Let us look for a moment at the vast army of administrators, site supervisors, skilled craftsmen and laborers who participated in the construction process. They were all under the control of a specially created Office for the Construction of Tōdai-ji, a government department presided over by a senior monk and high-ranking aristocrats of Emperor Shōmu's court. The office was divided into nine separate departments, each responsible for a different aspect of the project. A Timber-Collection Department dispatched lumbermen west to Harima, near Kobe on the Inland Sea, to obtain the forty-eight principal pillars, each thirty meters long and 1.5 meters in diameter, needed for the Great Buddha Hall. The forested mountains around Lake Biwa, north of Nara, provided the smaller timbers necessary to complete this and the many other worship and residential structures in the temple complex. A Transportation Department floated the timbers from mountain forests to collection points along local rivers, a hazardous operation at the best of times. The Building Department responsible for the prefabrication and assembly of the structures—the most labor-intensive aspect of the whole project—employed 227 site supervisors, 917 master builders and 1,483 laborers. At peak periods over 1,000 cooks prepared meals for the different craftsmen and laborers at the site.

The word *daiku* (lit., "great builder"), today commonly used for all carpenters, came into general use at this time to designate the chief builders in the Building Department. Immediately below them in rank were *shōku* (lit., "little builder") who helped supervise site activities. Below these were the ordinary carpenters who were referred to as *takumi* ("artisan"), or *mokkō* (woodworker). In time, the Chinese characters for "wood" 木 and "worker/builder" 工 were combined to create a distinctive Japanese character *moku* 杢 to describe the craft and profession of carpentry. The term is still found today in the name of the great traditional building firms such as Yasuimoku Kōmuten of Kyoto. The term *banshō* ("artisan on duty") also came into common parlance in the Nara period. It referred initially to carpenters who were sent to the capital from each region to participate in official building duties in a kind of corvée labor. Eventually the term was applied to carpenters in general until well into the medieval era. The origins of many of the terms which were building blocks of the language of Japanese architecture can thus be traced to this period. Although the precise usage of many of these terms has changed over the course of the many centuries since the Nara period, we cannot appreciate their later significance without first understanding their origins.

Along with organizing state projects, the Nara government made efforts to standardize weights and measures as part of the concentration of bureaucratic control over the nation. The city of Nara itself was laid out in a grid plan of standardized dimensions, and the process was carried over into the extensive rice fields of the capital region, which were reorganized into a similar orderly geometric pattern. All this required standardized measurements, but several *shaku* units of linear measure were actually in common currency. One was based on Korean measurements, the most direct point of Asian mainland contact. Another was based on Chinese measurements, from which the Korean measures were ultimately derived. Until the middle of the seventh century, the unit now known as the *Komajaku* (*shaku* of the northern Korean kingdom of Koguryo, pronounced in Japanese as Kōkuri or Koma) was in widespread use. It measured thirty-five to thirty-six centimeters and was used to determine the dimensions of some of the most important grave mounds. After the Taika Reforms of A.D. 646, in which many of the administrative practices of Tang government were adopted, the long-established Chinese system of a "small *shaku*" and a "large *shaku*," 1.2 times greater in size, was used at the official level. However, at least one Nara building, the early eighth-century pagoda of Kairyūō-ji, was still built using a *shaku* of 35.45 centimeters, effectively the *Komajaku*. Slightly earlier in 702, before the capital was moved to Nara, the government had adopted the Tang-dynasty practice of using a "large *shaku*" for land surveys and a "small *shaku*" for all other purposes. In 713 a law was issued which abolished the larger unit, stating that henceforth "each government ministry shall use the *shōshaku* [small *shaku*] for all purposes." This unit was approximately 29.5 centimeters in length, considerably shorter than the *Komajaku*. The *Ryōshūge*, a government commentary compiled in 822, explains that the relationship between the two units was fixed: "Five of the Korean *shaku* would henceforth be equal to six of the present day large *shaku*."[3] This "large *shaku*" (*tai-shaku*) became a general but still approximate standard.

This discussion of early Japanese measurements might be little more than speculation were it not for the existence of twenty-six Nara-period *shaku* rulers, some of them elaborately decorated with gold and mother-of-pearl, in the Shōsō-in, where the personal treasures of Emperor Shōmu were preserved after his death in 756. The *shaku* on these rulers have thirteen different lengths, ranging from 29.421 centimeters to 31.209 centimeters. Most fall into the 29.4 to 29.7 centimeters range. The earliest measuring ruler ever found was excavated at Hōryū-ji and is now preserved in the collection of the Tokyo National Museum. It is an elegantly decorated 1-*shaku* ruler designated as an Important Cultural Property because of its historical importance. The *shaku* of this ruler is 29.63 centimeters in length, conforming closely with the measurements of the rulers preserved at the Shōsō-in. It has not been proven, however, if this ruler was actually used for the rebuilding of the temple complex after it was gutted in the mid-seventh century fire or if it is a ceremonial object from a slightly later date.

It would seem, therefore, that government attempts to standardize measurements had little long-term impact on the realm of building practice in the Nara period. Despite official efforts at standardization, artisan workshops stubbornly retained their own units of measure. Extant buildings from the eighth to the eleventh centuries have *shaku* units of different lengths. The pagoda of Daigo-ji, one of the few surviving tenth-century buildings, even has *shaku* of different length for each of its five stories. For some, like the Konjiki-dō of Chūson-ji, completed in 1124, the *shaku* used is as much as 32.27 centimeters long, well over the length of the official standard of the Nara period.

COURT CULTURE AND THE DECLINE OF MONUMENTAL CONSTRUCTION

Few buildings survive to give witness to the long centuries of flourishing aristocratic culture known as the Heian period. The period was centered on the court of Heian-kyō, the city later known as Kyoto, which was established in 794 as the new capital in order to break the excessive influence of the powerful Buddhist sects over the state at Nara. Although the layout of Heian-kyō represented a modified continuation of Chinese principles of grid-plan order, the period to which it has given its name was an era of prolific indigenous creativity. After the fall of the Tang dynasty in China in the early tenth century, the Asian mainland lapsed into chronic instability lasting several centuries and contributed little more to Japan until the end of the twelfth century.

With the move from Nara to Heian-kyō, the emphasis shifted away from monumental state projects to more intimately scaled buildings constructed for the aristocratic families who dominated the court. The two most important buildings still standing, the Phoenix Hall of Byōdō-in at Uji to the south of Kyoto, and the Konjiki-dō of Chūson-ji at Hiraizumi in the far north of Honshu, are the products of direct patronage by the Fujiwara, the most powerful of these families. The Fujiwara dominated the Heian court for most of the tenth and eleventh centuries, and these delicately proportioned buildings, which still delight us with their infinitely detailed interior patterning in gold, silver, mother-of-pearl and semiprecious stones, bespeak the refined self-indulgence of the court. If these buildings are any guide to high cultural achievements in architecture in this period, it can be seen that the emphasis had shifted from structure to decoration. It was the skills of the ornamenting professions, rather than those of the master builder, that were now in greatest demand.

Shinden-zukuri, the residential form developed for the mansions of the aristocracy of Heian-kyō, had open-plan halls linked by covered galleries in an informal arrangment around extensive landscaped gardens. These buildings were roofed with cypress bark shingles (*hiwada-buki*) which reflected a quieter, native taste than the bombastic glazed terracotta tiling of Chinese inspiration used with such vigor in Nara. Bark shingles are also lighter than tiles, and complicated structural support and eaves bracket sets, like those seen on the Nara temples and halls of state, were unnecessary and indeed visually obtrusive.

Under these circumstances there was little change in the tools used for cutting and finishing timbers. The methods of preparing pillars and beams, of cutting mortises and tenons and of smoothing surfaces with adze and planing knife, remained firmly grounded in conventions established in the Nara period and probably stagnated because of the limited opportunities for large-scale work. The government of the Heian period did not maintain a large architectural establishment, although families like the Fujiwara had in their service considerable numbers of craftsmen, including professional makers of Buddhist images, such as the famed Jōchō who carved the Amida Buddha housed in gilded splendor at the Byōdō-in. Moreover, there were none of the attempts to standardize measurements that marked the Nara establishment. If a building like the five-story pagoda of Daigo-ji (fig. 121) is any indication, there was untroubled anarchy in the units of measure employed by the architectural workshops, a feature in common with medieval Europe.

There are few buildings remaining from the Heian period, and even fewer tools: not one is known to survive. The only direct evidence of tool type is furnished by written sources of which the *Wamyō ruijushō*, an encyclopedia compiled in 936, is the most useful. According to this compendium, a carpenter had a tool set consisting of two axes, one for cutting at right angles and the other for splitting with the grain. He also had an adze, planing knife, crosscut saw, an unspecified number of chisels, three gimlets, an iron-headed hammer, one large and one small mallet, a rope or string marker for laying out a site, a water level, the ubiquitous sumitsubo and a marking pen and carpenter's square. Gouges used for carving work, a cold chisel, draw knife and iron files used by smiths for making the tools themselves were also included in the list.

There is a notable dearth of saws and planes, tools without which the modern carpenter could not imagine executing the simplest tasks today. The list also reveals that there was little variety in chisels. It is remarkable that there are few differences in this list of tenth-century tools and those tools used two centuries before in the building of the Nara capital.

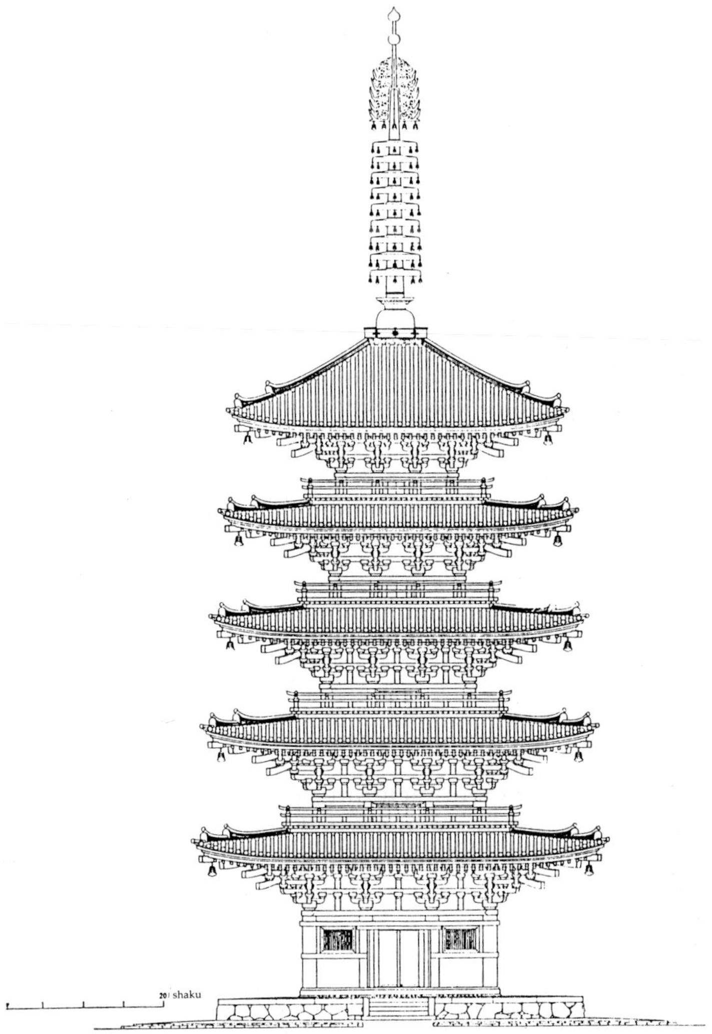

Fig. 121. Elevation of Daigo-ji Five-Story Pagoda, Kyoto. Ninth century. *Daigo-ji Five-Story Pagoda Restoration Report.*

The Liberation of Design—Invention of the Root Scale for the Carpenter's Square

Despite the general stasis in the development of tools of the hand in the Heian period, there was an important innovation made to the key tool of the mind, the sashigane. In the twelfth century, toward the end of this period, a square root scale was added to the rear, or *urame,* side of the measure's long arm. This scale set out the square root values of the units shown on the front of the same arm. Its deceptive simplicity makes it easy to overlook the significance of this innovation, which ranks among the most important of all inventions in the history of Japanese architectural design.

The urame scale transformed the carpenter's square into a versatile and powerful instrument of design. When the carpenter's square was first adopted from the Asian mainland, probably in the seventh century, it was merely a modular scale which could also be used for setting angles. The new scale added in the Heian period enabled the carpenter to perform two complex design operations with speed, precision and simplicity at any workshop or building site.

In the first of these operations, the scale could be used to determine the dimensions of a squared member that could be cut from a round log (fig. 25), an ingenious procedure already described in chapter 3. To state the obvious, square roots made logs easier to calculate. This improvement to the carpenter's square came as a result of the shift from round to squared pillars in residential architecture, one of the modifications to building styles which occurred as Japanese architecture matured in isolation from the continent. Round pillars were the norm in Chinese and Korean buildings, but the Japanese found that, when sliding screens became standard features, it made better sense to construct residences with square pillars. The interface of screen and squared pillar is neater and keeps out drafts more efficiently than that made with a round pillar. The creation of a "ready reckoner" for the basic tool of measurement indicates that the squaring of timbers had now become a daily concern.

The new root scale made another and even more significant contribution to building design. Easy calculation of curvature by describing parabolas based on square and cube roots became possible (fig. 23). The scale bestowed upon master carpenters the ability to design a variety of complex curved forms with relative simplicity. This marked a fundamental point of departure in the design of the roof, the dominant visual element in the Japanese building. The simple square root scale on the sashigane made calculation of curvature and placement of rafters and framing elements a matter of simple tangents. The most dramatic evidence of this development was the *kara-hafu*, a Japanese style of cusped gable (despite its name which perversely means "Chinese" or "Tang" gable). Its alternating concave and convex curves required perfection in the symmetrical placement of rafters and ribbing (fig. 122). By the twelfth century it was to be found adorning the side eaves of aristocratic gateways and the front barge boards of court carriages. Thus the Heian period ended with one of the simplest but most important developments in Japanese tool technology, one which laid the foundation for many of the design practices of later tradition.

DEVELOPMENT IN THE MIDDLE AGES OF STYLE, STEEL AND SAWS

The era in Japanese history known loosely as the "middle ages" embraces both the Kamakura (1185–1333) and Muromachi (1333–1573) periods. Likened by scholars to the era of feudalism in Europe, it was a crucial time in the development of new types of tools in Japan. At the beginning of the middle

Fig. 122. *Top,* Shōryō-in, Hōryū-ji, Nara, the earliest extant *kara-hafu* (cusped gable), dating to rebuilding undertaken in 1335 but reflecting the style of the gable built when this subprecinct was established in 1121; *bottom left,* Hirakara-mon, gateway with cusped side gables, of fourteenth-century Gyokuhō-in Subprecinct, Myōshin-ji, Kyoto; *bottom right,* late sixteenth-century Chokushi-mon (Imperial Messenger Gateway), Sambō-in Subprecinct, Daigo-ji, Kyoto. *Hōryū-ji Shōryō-in Restoration Report; Kokuhō jūyōbunkazai jissoku zushū: Myōshin-ji Gyokuhō-in; Kokuhō jūyōbunkazai jissoku zushū: Daigo-ji Sambō-in.*

ages, architectural activity had shrunk in both scale and frequency as the authority of aristocratic government, which was centered in Kyoto, withered away with the rise to power of the warrior class in the provinces. At the end of the period, society, government and culture had all undergone profound change. A succession of warrior governments, or shogunates, had employed architectural forms as an expression of their Zen-inspired ideals and as a setting for public audiences and private functions. The first shogunate, under the Minamoto and Hōjō families, founded a cluster of Zen monasteries around its headquarters at Kamakura, near present-day Tokyo. These monasteries were the focus for the adaptation of the architectural styles of Song-dynasty China to Japanese circumstances (fig. 9). The second shogunate of the Ashikaga family, for reasons of political expediency, reestablished the center of government in the imperial capital of Kyoto. Trade connections with China, which had lapsed for several centuries, were promoted with the new Ming dynasty, and treasures of refined lacquerware, ceramics and ink paintings poured into the shogunal courts at Higashiyama and Kitayama and into the major monasteries like Shōkoku-ji and Daitoku-ji. This mingling of warrior and aristocratic traditions gave Japanese culture some of its most exquisite and distinctive architectural expression. Tracing these changes through the design and building construction of the Kamakura and Muromachi periods uncovers the central importance of tool innovations to the creation of the new styles.

Song Stylistic Influence and the Mechanics of Assimilation

The power of the samurai, coupled with fresh influences from a China temporarily calm and reunited under the Song dynasty (960–1279), created dramatic changes in architectural style in the thirteenth and fourteenth centuries. The gaunt, monumental style of southern Chinese coastal areas had been tried out at the rebuilding of Tōdai-ji and related projects in the later twelfth and early thirteenth centuries, but the architecture of the Northern Song dynasty became the new orthodoxy favored by the warrior government for its Zen temples. These were a race apart from the earlier temple buildings, speaking a different language of elegance and proportion. The roofs had sweeping eave lines with graceful radial raftering and profuse clusters of multiple-arm bracket sets. Penetrating tie-beams (*nuki*) all but replaced the external tie-beams (nageshi) except as decorative affectation (fig. 123). The most sublime accomplishment to survive from the era is the pagoda of Anraku-ji in Nagano Prefecture (figs. 124 and 125). With its octagonal-shaped roofs delicately uplifted at each corner, this building reveals that elegant proportions were achieved through stronger framing and joinery techniques.

Although the Song style was used primarily for Zen temples, giving rise to the Japanese name *Zenshū-yō* ("Zen-sect style"), its influence was not confined solely to the religious sphere. It set new standards of taste for secular palaces and residences, where its curved forms proved as much of a challenge for the Way of the Carpenter as Zen was for the Way of the Warrior. The master builders responsible for creating Kennin-ji in Kyoto, the first Song-inspired building complex completed in the early thirteenth century, and Kenchō-ji, finished in the

Fig. 123. Penetrating tie-beam (*nuki*) of Kiyomizu-dera Three-Story Pagoda. Early seventeenth century.

Fig. 124. Elevation of pagoda of Anraku-ji, Nagano Prefecture. *Anraku-ji Restoration Report; Nagano-ken no kokuhō jūyōbunkazai kenzōbutsu shūri kōji hōkokusho* 1.

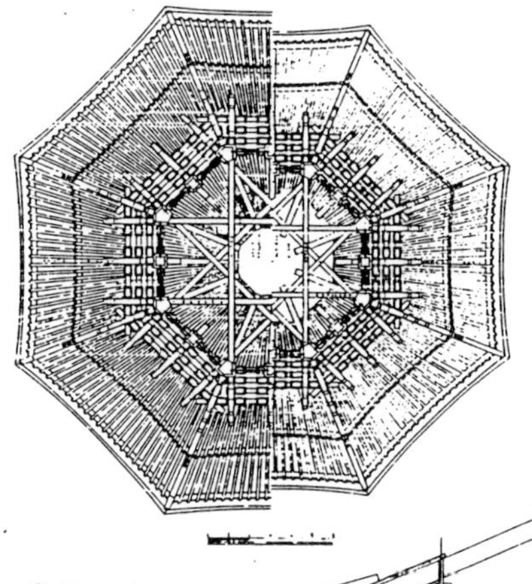

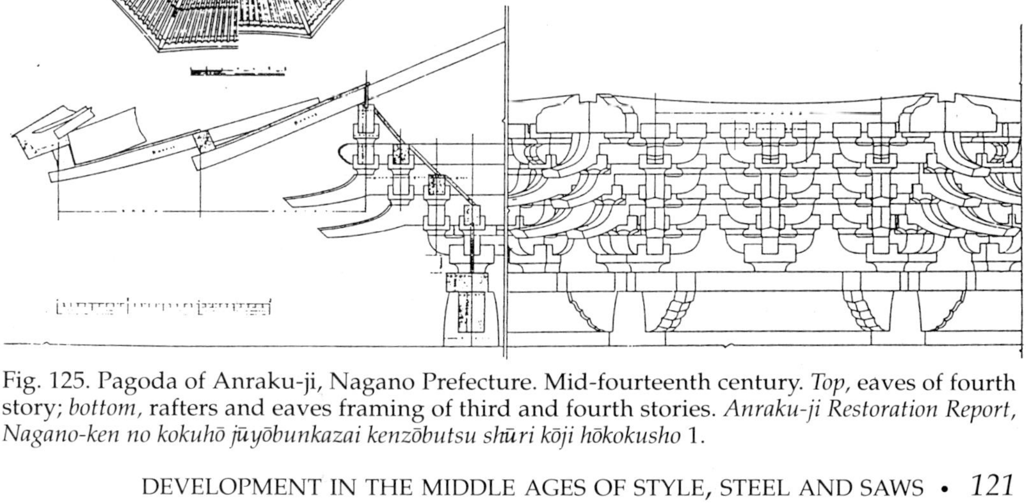

Fig. 125. Pagoda of Anraku-ji, Nagano Prefecture. Mid-fourteenth century. *Top,* eaves of fourth story; *bottom,* rafters and eaves framing of third and fourth stories. *Anraku-ji Restoration Report, Nagano-ken no kokuhō jūyōbunkazai kenzōbutsu shūri kōji hōkokusho* 1.

city of Kamakura around 1253, were forced to improvise. They had to reproduce the new style without complete knowledge of Song building techniques because information was transmitted by monks and traders who had no specialist knowledge of the ways and means by which such buildings were made in China. To emulate the curvilinear extravagances of Song architecture, carpenters cut deep mortises through principal pillars to house-penetrating ties, without knowledge of the techniques or rationale for effecting these changes. It is no wonder that many of the surviving examples of religious establishments founded in the thirteenth and fourteenth centuries are delightfully eclectic in style but decidedly conservative in method of construction. Many of them were a case of a dash of Nara temple style mixed with some Song inspiration and a great deal of homegrown architectural wisdom.

One important outcome of all these pressures for change in building style created by foreign influence was increased experimentation with the carpenter's square as an instrument of design. The design potential of tangents and triangulation, parabolas and polygons, was realized by the carpenters as they met the new cross-cultural challenge. It could be said that in the Heian period carpenters learned to walk with their sashigane scale; in the Kamakura period, as the Song stimulus increased the pace of experimentation, they learned to run. By the Muromachi period they were flying, building with a lightness that lifted the roof from all apparent earthly constraints. Now the placid kara-hafu of Heian, with its crouched stance, had achieved a soaring silhouette close to pure fantasy. The eaves flared away from the high arched apex in a series of effortless sweeps, concave curve blending into convex slope, rafters beneath accentuating the flow of the gable with a gentle syncopated rhythm (fig. 122). This truly represents perfect harmony between the instruments of the mind and the tools of the hand.

Pictorial Evidence of Tool Techniques in the Kamakura Period

An accurate and lively answer to the question "what did the building site of a temple like Anraku-ji or Kennin-ji look like in the thirteenth and fourteenth centuries?" can be found in the eye-witness accounts of narrative picture scrolls (*emakimono*). These scrolls, frequently of considerable length, first became an important form of artistic expression during the Heian period. In the new, vigorous warrior age, their focus shifted from the rarified life and loves of the imperial court to scenes of everyday activities, including glimpses of construction sites. The most detailed scenes are to be found in the *Kasuga gongen genki-e* and the *Matsuzaki tenjin engi emaki* (pls. 2–4). Both paintings, reliably dated to 1309 and 1311 respectively, are well preserved and depict carpenters at work. There is an eye for human interest as well as a delight in detail which makes them convincing documentary evidence of tool type and usage.

The *Kasuga gongen genki-e*—an alternative title is *Kasuga gongen reigen-ki*—(pl. 2) tells the history of the Kasuga Shrine in Nara. The first scroll shows the construction of the Chikurin-den in 948 but, fortunately for our understanding of the Kamakura period, the tools and techniques depicted belong to the artist's own time. All the tools and activities are painstakingly drawn and readily

Fig. 126. Detail of shrine building showing use of planing knives (*yari-ganna*). Detail of pl. 2, *Kasuga gongen genki-e*, scroll 1, 1309.

recognizable. The site is a hive of industry; carpenters are stripped to the waist as evidence of their physical exertions. They are laying out the site with string markers or carefully prefabricating the framing timbers. At center scene stands the tōryō, clad in loose pantaloons, his kensao (long measuring ruler) resting on one shoulder as he directs operations. Nearby we can see a carpenter checking the level of the foundations as an apprentice pours water into the water level. A team of three carpenters energetically pound a foundation stone into its correct position on the podium. At the adjacent workshop the timbers are being split with wedges, marked with a snap line, and measured for roundness with the urame scale on the back of the carpenter's square. Beside this carpenter sits the sumitsubo, ready for instant use. Close by, groups of carpenters skillfully smooth rough timbers with adzes, while the long-handled planing knives, yari-ganna, are pulled or pushed across the surface of the wood to give it a smooth finish (fig. 126). We can appreciate the difficulty in operating these planing knives from the studied concentration and firm, two-handed grip of the carpenters. It is hard to imagine a tool which demands better eye-hand coordination. It helps explain the preference of Japanese carpenters for sitting down on the job. A seated position gives better control for precision work and permits concentrated activity for sustained periods of time. This would have been essential for all the painstaking cutting and smoothing operations depicted in medieval scrolls. Sitting rather than standing gives the arm a more stable pivot point, important in avoiding gouging the surface of the wood when using the planing knife. This same sitting position also leads naturally to a pull-action for the saw and plane, as we have seen in Chapter 3.

The second narrative scroll which shows construction activities is the *Matsuzaki tenjin engi emaki*, painted in 1311, just two years after the Kasuga scene. Scrolls 4 and 6 in this series include scenes of frame-erection which, conveniently for us, is the next phase in construction (pls. 3 and 4). The teamwork necessary to assemble the prefabricated parts is seen in the closely coordinated groups of craftsmen struggling to position the heavy timbers of the upper framing. In Scroll 4 a carpenter, eager to assist his struggling colleagues, clambers up the narrow timber scaffolding, a long step-bladed trowel chisel in his hand ready to fix a mortise which is not housing its tenon properly. The tōryō directs activities from below, his long ruler ready to check the vertical dimensions of the framework. In the other scene of building activities in Scroll 6, we find the tōryō using his sumitsubo to make sure the frame is plumb. Surveying the scene a short distance away sits the patron of the project, the ill-fated courtier Sugawara no Michizane, dignified in demeanor and dress and no doubt gratified by the special efforts made by the carpenters for his benefit. We learn a great deal in these scrolls about tools and their practice in the hands of the men who use them.

Development of the Pull-Action Crosscut Saw

These scenes of building activity in Kamakura period paintings also provide important evidence that, by the early fourteenth century, the Japanese had improved the design of the Chinese crosscut saw into a leaf-shape blade and had altered the cutting action from a thrusting to a pull-stroke. Both these developments were a turning point in the evolution of saws in Japan.

The new, leaf-bladed saw is shown in most scenes of construction activity of the thirteenth and fourteenth centuries, including the Kasuga and Matsuzaki scrolls. The artists of this era took particular delight in emphasizing its existence; they accord it the visual prominence of an exciting technological innovation. In a scene showing the building of the Ishiyama-dera, painted about fifteen years after the Kasuga and Matsuzaki scrolls, a carpenter waves the leaf-saw authoritatively to emphasize his point of view in an argument with a colleague. Meanwhile, in the foreground, another carpenter has a leaf-saw tucked jauntily into the waistband of his trousers. In the Kasuga Shrine construction scene, the leaf-saw occupies pride of place in a special rack on the lid of a carpenter's toolbox, while another such saw is gripped grimly in the teeth of a carpenter as he hauls a heavy beam across the work site.

A leaf-shaped blade has a more efficient action than a straight-edged saw because of its breasted cutting edge. Nonetheless, the performance of this new tool was limited. A ceiling panel shown in the Matsuzaki scrolls is the largest piece of timber, in any of the Kamakura-period narrative scrolls, which this type of saw is seen cutting. Judging from the comparative size of the carpenter doing the work, the panel can be judged to be about thirty centimeters wide and three centimeters thick. It is significant that one carpenter in the Matsuzaki scrolls is using a leaf-saw with the sawdust cascading toward him. Covering the carpenter with sawdust is the proud prerogative of the saw which cuts on the pull stroke, one of its few disadvantages over the pushed saw in which the dust exits the kerf on the far side of the timber. It seems, therefore, from careful examination of these scenes of carpenters at work that the sitting position led naturally to a pulling action for cutting tools as well as some subtle improvements in their design.

We are not left to speculate about the nature of this new saw on the evidence of paintings alone, for two leaf-saws have been discovered at archaeological excavations of medieval building sites. In 1980, a saw blade close in appearance to that of the leaf-saws shown in the Kasuga and Matsuzaki scrolls was discovered at Kusadosengen-chō, in the city of Fukuyama, Hiroshima Prefecture (fig. 127). The remains of an entire medieval town were exposed in the local riverbed after it changed course following a major flood in 1919. Systematic archaelogical excavations began in 1961 and have since revealed pillar stumps belonging to residences and shops as well as a plethora of objects from everyday life, including ceramic bowls, wooden tubs, fragments of clothing and thousands of coins.

The saw was excavated in rusted fragments, but X-rays showed clearly its original shape and the precise configuration of the teeth. Reconstruction of the appearance of the saw reveals that it originally had a long, narrow blade 40.65 centimeters in length and 4.3 centimeters at the widest part (fig. 127). It was similar in appearance to the leaf-saws shown in the Kasuga and Matsuzaki scrolls, so it is safe to infer that it dates to the Kamakura period, especially as the township from which it was recovered also dates to this period. The X-rays also revealed that the teeth originally radiated in a continuous arc around the convex, curved cutting edge. Experiments carried out in 1981 at the research laboratory of Gongō-ji demonstrated that this saw was not an efficient cutting tool and that it made little difference whether it was pushed or pulled to cut the wood, a very interesting comment on the state of development of the saw nearly seven hundred years ago. The distinctive, pull-action saw was still a few years in the future.

The second leaf-shaped saw was also discovered in 1980, at the Shimokōri site of a Muromachi-period building in Ueno, Mie Prefecture. The blade is in remarkably good condition and reveals that, in the hundred years that had elapsed from the time of the Kusadosengen-chō saw, a significant improvement in cutting efficiency had been achieved by redesigning the cutting edge (fig. 128). The teeth of the Shimokōri blade point backwards toward the handle of the saw, indicating that it could only have been used with a pulling action. It is basically a crosscut saw but would have worked reasonably well for some small ripsaw jobs such as fashioning tenons. Several generations of experimentation and development by carpenters and smiths are evident in the differences between these two blades, clearly and reliably documenting two of the major stages in the development of the Japanese crosscut saw.

The second saw is doubly significant because of the way in which it was made. It consists of high-quality laminated steel, a clear indication of the advances in steel-making technology in medieval Japan. Recently, Miyano Tetsunosuke II of Miki, a master sawsmith of Mie Prefecture, was commissioned to reconstruct the process by which this saw was made. The eighty-three-year-old master smith, assisted by three senior apprentices, took five days of intensive labor to re-create the process by which the saw was made in the Muromachi period. The long and sophisticated procedure entailed repeated heating of the steel, shaping on an anvil and quenching in water to give the cutting edge its brittle hardness. To give the blade the springiness necessary to perform well in the hands of a carpenter, when the forging was nearly completed the steel was coated with clay and straw and baked slowly in the forge to temper the blade by heating from the core outwards.

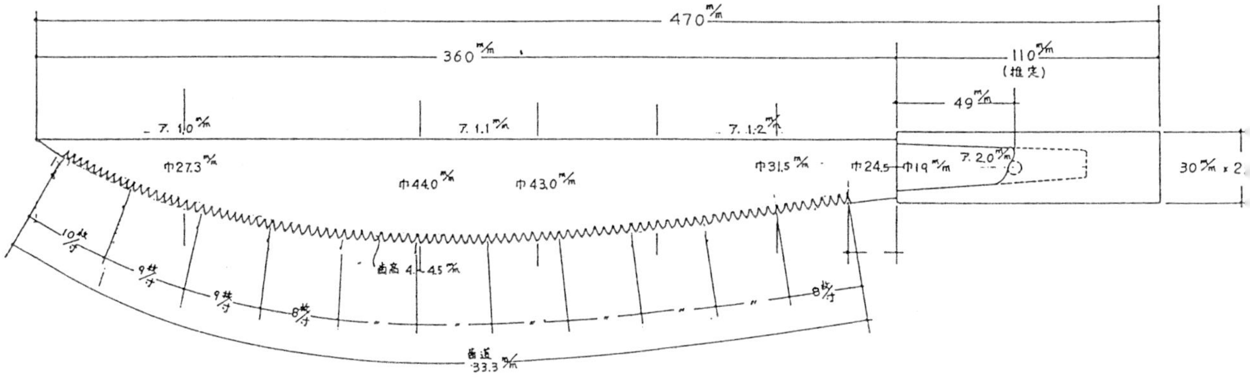

Fig. 127. Twelfth- or thirteenth-century leaf-blade crosscut saw excavated at Kusadosengenchō, Hiroshima Prefecture. The teeth radiate in a continuous arc and were used with either a push or a pulling action. Reconstruction drawing based on X-ray photographs.

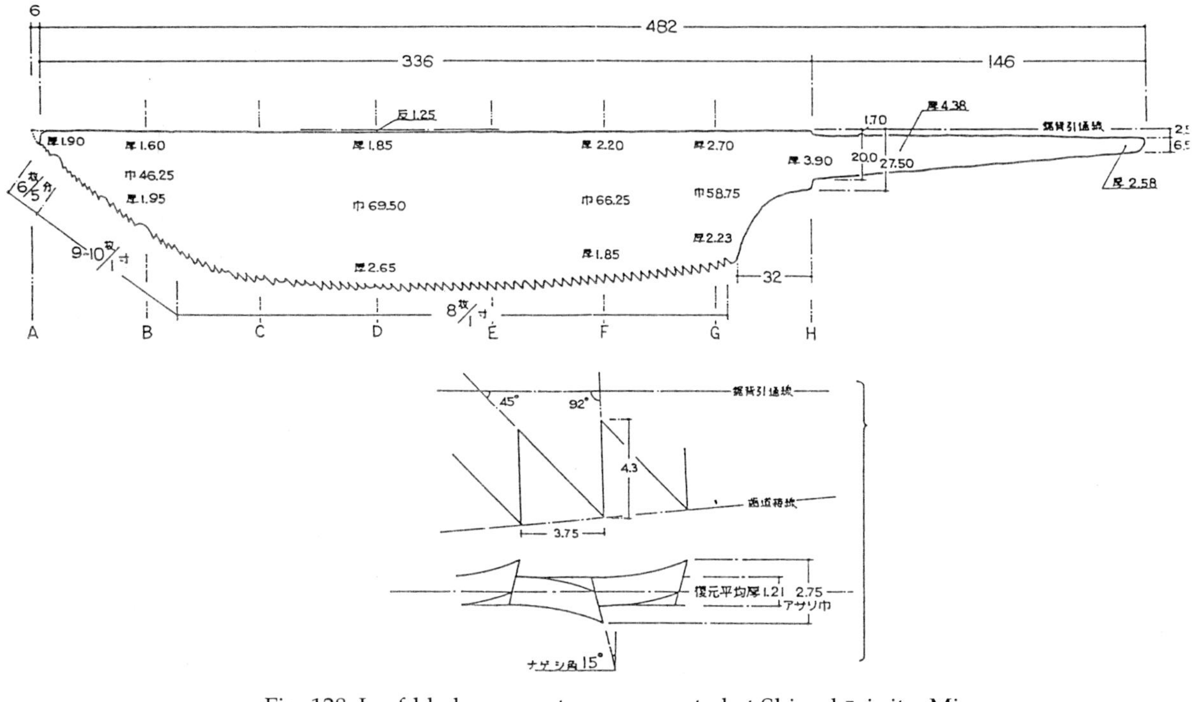

Fig. 128. Leaf-blade crosscut saw excavated at Shimokōri site, Mie Prefecture. Ca. fourteenth-century. The teeth point toward the handle, indicating that it was designed to cut on the pull-stroke.

This saw, then, based on improved steel technology and adapted to Japanese work habits, is far removed from the Chinese push-action "grinding saw" used at Hōryū-ji and Nara. It appeared at a time of intensified building activity in Japan under the stimulus of warrior patronage. It was used by the carpenter in a sitting position, and its teeth were configured with their cutting edges toward the handle to permit a well-controlled pulling action.

The pull-action crosscut saw of the Kamakura and Muromachi periods was to become the prototype for the general roughing saw, the hanamaru-noko, and for more refined crosscut saws of the sixteenth and seventeenth centuries.

The Medieval Construction Crisis

The Kasuga and Matsuzaki scrolls are of immeasurable help in capturing the state of the art of building in Japan about the year 1300. This date is significant because, with the advantage of historical hindsight, it marks the approximate end of an era in the history of Japanese tools. By the time the children playing near the toolbox at the shrine construction site had passed on their carpentry skills to their own children, changes had occurred which forever altered the character of cutting tools, the nature of architectural form, and the organization of the building professions themselves.

These radical changes were precipitated by the medieval equivalent of the modern-day energy crisis: the Japanese began to run out of straight-grained timber of structural dimensions. The increased demand for buildings throughout the Kamakura period created a severe drain on available material resources. Supplies of straight-grained timber, particularly hinoki, dwindled alarmingly. During the seventh century, which had been the busiest period of temple construction in the Asuka-Nara region, supplies of good timber had been readily available in the forest-clad surrounding mountains. However, by the time of the construction of Tōdai-ji in the mid-eighth century, most of the timbers had to be carried from the mountains around Lake Biwa, well to the north. The giant pillars for the Great Buddha Hall were brought by sea and river over five hundred kilometers from what is now the Kobe region. The rebuilding of Tōdai-ji at the turn of the twelfth century was actually delayed for several years, in part because of difficulty in locating replacement pillars. Timber was eventually located in the province of Suō, near the extreme Western tip of Honshu.

The case of Tōdai-ji typifies the general circumstances prevailing at the end of the Kamakura period. Unlike the situation in the twentieth century, when the Japanese have been able to import Taiwanese cypress in order to alleviate their wood shortages, the carpenters of medieval Japan had no alternative but to change their work practices. They now had to make more joints to connect smaller timbers and to use the capriciously grained keyaki, a type of timber which cannot be split in a straight line, for structural needs. This rendered the traditional technique of splitting and smoothing timbers useless. The shortage of wood compounded the pressures on the carpentry profession created by the greater use of penetrating tie beams as part of the Song-influenced style. New and improved tools were urgently needed, both for better joinery and for longitudinal cutting of lumber.

The Invention of Steel-tipped Cutting Tools

The development of steel-tipped cutting blades was one of the great technological innovations in architectural infrastructures between the end of the

Heian and the modern period. The crisis created by the difficult Song styles and the shortage of straight-grained timbers stimulated the application of techniques of steel making to carpenters' tools. Steel blades were infinitely sharper and more durable than the wrought-iron tools they replaced.

In addition to the technique of laminating steel to make saw blades, the Japanese perfected the art of making chisel blades, and later plane blades, by adding a cap of high-carbon steel to a malleable iron core by repeated heating, beating and quenching (pl. 9). The flexibility of blades was further enhanced by tempering with clay, leaving the cutting edge exposed to give it a tougher temper from quenching. The same technique had been perfected in the making of samurai swords.

This method of making steel-tipped blades is shown in the *Shokunin zukushi-e,* a series of paintings dating to about 1600 in the collection of the Kita-in, Saitama Prefecture, which show artisans at work (fig. 129). A master smith crouches on a coiled straw mat before the glowing coals of his forge, surrounded by the orderly chaos of the tools of his trade. There are steel-headed hammers with handles of different lengths for shaping the white-hot steel and pincers for handling his work. On the wall beside the forge are hung the draw knives (*sen*) used for smoothing the surface of the steel to a fine finish, and heavy-bladed, short-handled choppers for cutting the steel to length to make the layers for lamination. The smith works with an apprentice or fully trained assistant, his wife keeping an observant eye on activities while busying herself with household duties.

The smith shown in this seventeeth-century scene has a fixed place of work, his rustic cottage. For much of the medieval period, however, smiths were itinerant, following builders to each project and setting up shop at the construction site. There they would labor from dawn to dusk, not only making and repairing tools in close consultation with the users but also manufacturing the multitude of nails, clamps and other metal fittings needed for finishing a building. By the sixteenth century, smithing as a trade had stabilized in key centers such as Kyoto, and increasing specialization followed naturally. Separate subprofessions for the makers of saws, chisels and plane blades emerged. The smith shown in the *Shokunin zukushi-e* has drawings of chisels and plane blades fixed to the wall behind his forge for ready reference, an indication of his personal speciality.

Chisels were the tools most immediately affected by the new technology. They were, to strain the popular expression, on the cutting edge of building techniques. Steel-tipped chisel blades gave carpenters new power to cut the penetrating ties needed for Song-inspired buildings. They were also required to make the stronger spliced joints necessary to connect shorter timbers now used to compensate for wood shortages. More types of chisels were developed to enhance mechanical efficiency. The trowel chisel, for example, first appears in the Matsuzaki scroll.

Two other changes to basic chisel design added further to their cutting efficiency. First, double-beveled edges were replaced with single bevels. Second, the tang was added to reinforce the connection between blade and wooden handle. Both changes were made against a backdrop of blistered palms and

Fig. 129. *Shokunin zukushi-e* with a detail of a smith (*kajiya*). *Kita-in. Saitama Prefecture.*

broken chisels as carpenters struggled with unsuitable woods to create the fashionable designs of the new age.

The change to the single bevel was a subtle but significant improvement. Prior to this, chisels had been used both for making joints and, in conjunction with oak or iron wedges, for splitting lumber to make pillars and beams. The cutting edge of these chisels was beveled on both sides to facilitate penetration through the grain, a feature to be found on the tomb chisels of the fifth century. With the abandonment of cleaving as a longitudinal preparation technique in the late Kamakura and Muromachi periods, the chisel became the specialized instrument for joinery. A double bevel proved inaccurate for precision work; it is easier to cut a clean mortise or tenon with a single-bevel chisel because the flat steel side of the head can be placed flush against the surface of the wood as it penetrates the grain.

The second of these changes occurred about the same time. The addition of the tang made it possible to apply considerably greater force to the chisel as it cut through temperamental woods. Henceforth in Japan the tanged and socketed form was the standard style of chisel. It is shown in the working drawing of the tool smith in the *Shokunin zukushi-e* of the seventeeth century and, interesting to note, has changed little down to the present day.

What was the impact of the more efficient chisels on building techniques? There was a marked change in the gooseneck mortise-and-tenon joint (*kama-tsugi*), the most important splicing joint used for connecting timbers longitudinally. Extant buildings which have been dismantled for restoration work reveal a clear change in the design of this joint in the late fourteenth and fifteenth centuries. The earliest known use of a gooseneck type of mortise-tenon is for beams and ground sills of the "lean-to" roofs added to the first story of the five-story pagoda and Kondō of Hōryū-ji in the eighth century (fig. 130.a). These gooseneck tenons are stubby in proportions, and the heads of the tenons are squared like a hammerhead. Although an oblique-sided tenon head similar to the later gooseneck tenons was used on a few buildings such as the mid-eighth century Dempō-dō of Hōryū-ji and the tenth-century pagoda of Daigo-ji, the thickly proportioned rectilinear mortise and tenon remained standard until the fourteenth century. In the twelfth century there had been some improvements in the design of these joints with the incorporation of lapped joints into the hammerhead for greater strength, as seen at the Main Hall (Hondō) of Ishiyama-dera (fig. 130.b). However, most gooseneck tenons made up to the fourteenth century avoid curved and angled parts in favor of simple planes set at right angles. It is simply easier to work with the grain in this manner.

The form of the gooseneck tenon changed considerably during the later part of the fourteenth century. The hammerhead joint became the exception rather than the rule as a variety of sophisticated oblique-sided tenon heads came into use. Some of the earliest known examples were used for the penetrating head ties of the Niō-mon of Tōfuku-ji (fig. 130.c), built in 1391 under the patronage of the Ashikaga shogunate as one of the new centers of Zen Buddhism. By the later part of the fifteenth century, gooseneck tenons had become even more sophisticated. The purlins and penetrating head ties of the Main Hall of the Kita Muro-in, a subprecinct of Hōryū-ji, have detached gooseneck tenons added to stub tenons to provide additional structural strength (fig. 130.d). Later joinery was to become even more complex, moving from the sphere of structural necessity to the realm of self-indulgent art.

Adoption of the Two-man Frame-saw

The crisis of lumber supplies was finally resolved by the introduction from China of a two-man frame-saw (*oga*) for ripsawing lumber (pl. 5). Its use in Japan is first mentioned in documents describing building work carried out at the temple of Tō-ji in Kyoto in the 1420s, about a century after the Kasuga and

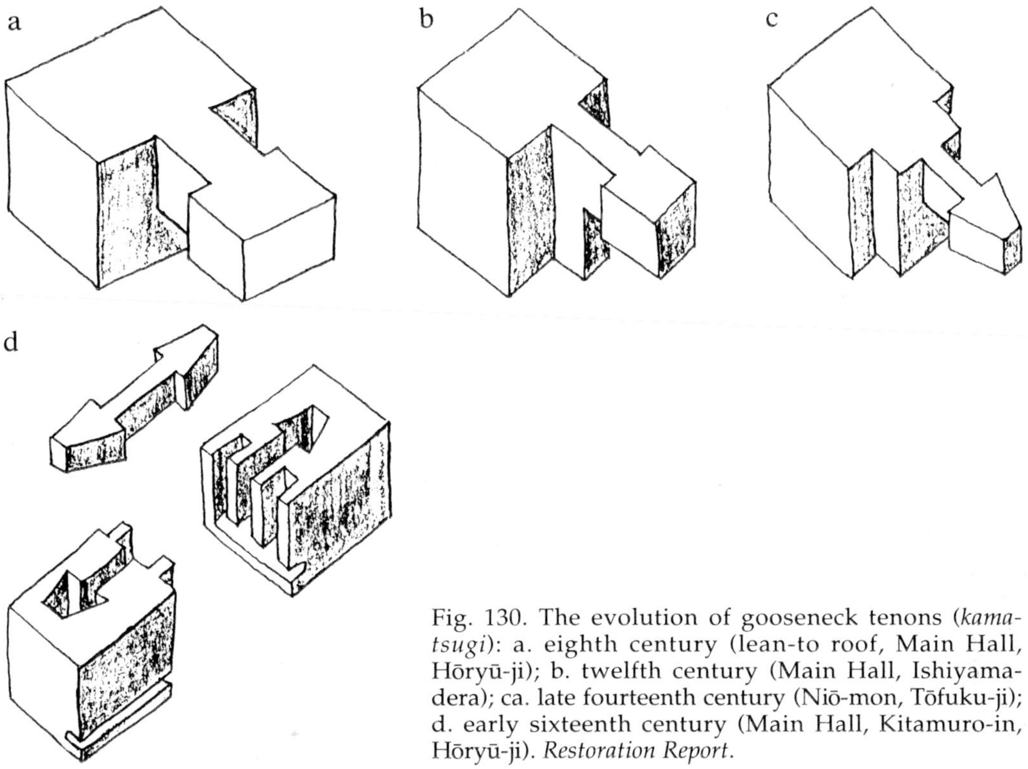

Fig. 130. The evolution of gooseneck tenons (*kama-tsugi*): a. eighth century (lean-to roof, Main Hall, Hōryū-ji); b. twelfth century (Main Hall, Ishiyama-dera); ca. late fourteenth century (Niō-mon, Tōfuku-ji); d. early sixteenth century (Main Hall, Kitamuro-in, Hōryū-ji). *Restoration Report.*

Matsuzaki building scenes were painted. By the middle of the fifteenth century, this saw is frequently noted in contemporary written records.

The Japanese had probably been aware of the existence of the frame-saw by the twelfth century, but plentiful supplies of straight-grained timber suitable for longitudinal splitting had made the use of a large ripsaw unnecessary. There was no need to saw laboriously through a piece of timber which could be split quickly and cleanly. As noted earlier, split timbers are also free from the warping endemic to wood prepared by sawing. With the depletion of timber stock in the medieval period, however, the frame-saw came into its own. It was now possible to cut a long piece of ill-tempered, convoluted grain lumber into cleanly finished structural timbers or neat, thin boards for use as flooring or wall and ceiling panels. It also prompted the creation of one of the most distinctive and beautiful Japanese decorative arts—paintings done directly on wide cedar panels cut with the frame-saw. Depictions of birds and flowers, mythological creatures and exotic animals began gracing the corridors of many palaces and shrines, captivating their viewers with bold color set against the clear, lively grain of the natural wood background.

The frame-saw used in Japan consisted of a steel blade from one to two meters in length mounted on a light, H-shaped bamboo frame. The blade was tightened by an oiled hemp rope wound tightly with a piece of wood in the manner of a tourniquet to stretch the lower part of the frame. This device kept the blade under tension. This tension was good for the nerves of the sawyer, because it increased the strength of the blade in the same way as the pulling action of later single-person saws increases their efficiency. As the blade became loose and ineffective with heavy use, the twisted rope served to restore the tension needed for the frame-saw.

The frame-saw is a familiar tool because it had long been in use on the Asian mainland, possibly tracing its roots along the Silk Route back to the Roman Empire. Saws of similar type may be seen in Roman frescoes, although tensioning devices seem to have been an invention of medieval Europe.

The frame-saw was extremely arduous to use, as may be imagined from the efforts of the two men immortalized in the *Sanjūniban shokunin uta-awase* (pl. 5). This set of illustrations of thirty-two different professions painted at the turn of the sixteenth century furnishes important documentary evidence of a change in the organization of the building industry which was prompted by the widespread adoption of the frame-saw for ripsawing.

Until the fifteenth century, there was only one main division of labor in the organization of the timber-building profession. The task of felling and rough hewing the lumber belonged to timber cutters who lived and worked in the mountain regions where the best supplies of timber were to be found; the carpenters themselves carried out all the arduous work of hewing and splitting the logs before they began detailed marking and finishing.

The advent of the ripsaw in Japan called into existence a new profession, the *oga-hiki*, or "frame-saw cutter," later known as *kobiki*, or simply "sawyer." The first reference to this profession appears in written records of 1443 concerning the famous temples on Mount Kōya. The scene from the painting of the thirty-two professions establishes that by about 1500 the sawyers ranked along with carpenters and stonemasons as one of the major building professions. As the *Sanjūniban shokunin uta-awase* scene reveals, sawyers worked in pairs, the master sawyer positioned on top of the piece of lumber to direct the cutting action while his assistant toiled away from below.

The operation was similar to methods employed by the pit-sawyers of Europe and America until a generation ago and may still be seen in remoter parts of China and Nepal today. Indeed, one could be forgiven for mistaking Walter Rose's description of sawyers at work in England at the beginning of the twentieth century with a first-hand account of the Japanese team at work in the Muromachi period painting. "How fascinating it was to watch them at work," Rose wrote. "The ceaseless up and down movement of the large shining saw wielded by two strong men at the pit; the slow creeping forward of the cut (in a large log so slow as to be scarcely perceptible) yet ever advancing towards the toes of the man standing on the log—though he always moved them back just in time; the completion of the log, when only a short length of each cut remained to be done and board after board fell off side-ways, each beautiful with grain markings that eye of man had never before seen."[4] It seems that the creation and discipline of the rip-sawyer in building practice has been a universal experience.

The establishment of the new profession of the ripsawyer in the fifteenth century is of great significance both to the organization of the building

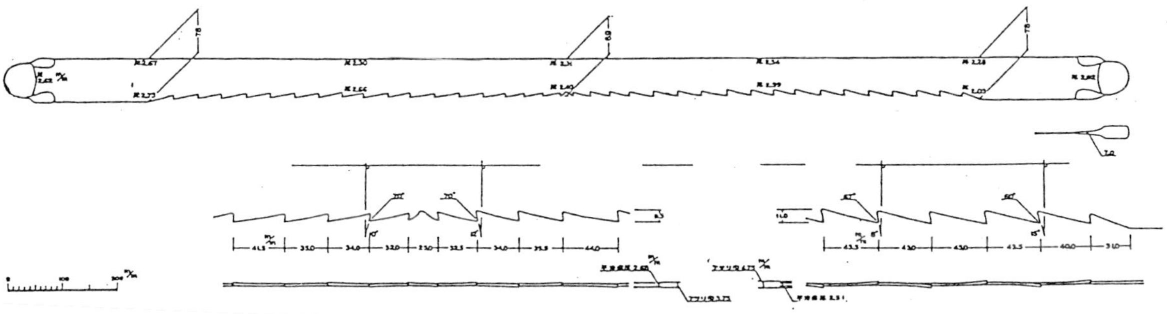

Fig. 131. Blade of an early fifteenth-century two-man frame-saw. Shakubu-ji.

professions and to the nature of the construction process itself. It speeded up the whole building process by the simple expedient of division of labor, and by freeing the carpenters from the most grueling part of timber preparation, doubtless left them with more time and energy to concentrate on the perfection of the details of their craft.

At least eighteen frame-saw blades from this era are known to have survived. The oldest is thought to be the blade preserved in a storehouse of Shakubu-ji, a temple in Kobe (fig. 131). It seems to have been employed for the construction of a Yakushi Hall (Yakushi-dō) and three-story pagoda in the early Muromachi period; in which case, it may be dated to slightly earlier than the first mention of the existence of frame-saws in documents of the 1420s.

An examination of the technical features of this particular blade is rewarding, for it reveals detailed information about forging technology and sawing techniques in the early days of the history of the frame-saw in Japan. The blade is 2.056 meters in length to the tips of its securing rings. The steel appears to have been cooled after forging by rapid quenching in water to give it a harder edge than that possible by slow annealing with a clay mixture. The teeth are sharply raked with a pronounced set of up to 5 millimeters to each side. Their most distinctive feature is that they are arranged facing away from the midpoint of the blade in each direction; thus, the saw had to travel only half the length of the cutting edge (approximately 72 centimeters) on each stroke for maximum cutting efficiency. The splayed tooth configuration seems to have been the standard style for the two-man frame-saw, in keeping with the Chinese prototype upon which it was based. Useful information about the oga is provided by the *Wakan sansaizu-e* published in 1712. This title may be translated as the "Japanese-Chinese Illustrated Assemblage of the Three Components of the Universe," explaining its modest aspirations as a world encyclopedia modeled on Ming dynasty compendia. Needless to say, it includes a section on tools.[5] This confirms the supposition about splayed tooth configuration when it states that "the oga is 6 *shaku* in length, the teeth going in opposite directions from the center [of the blade]" (fig. 132).

Curiously, all the teeth of the oga depicted in the *Sanjūniban shokunin uta-*

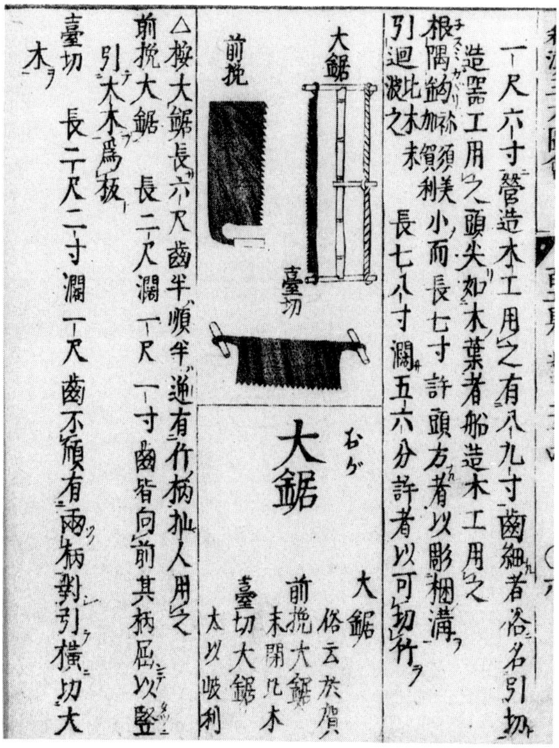

Fig. 132. Section of *Wakan sansaizu-e* (1712) describing saws. *Right*, the two-man frame-saw (*oga*) beside the wide-bladed ripsaw (*maebiki-oga*); *below*, the crosscut lumber saw (*daigiri*).

awase are pointing in the one direction. Previously, experts conjectured that the artist had made a mistake, but an actual example of an oga with teeth pointing in one direction as shown in the painting has now been discovered. It is preserved in the Kimura Ethnology Museum in Imari City, Saga Prefecture, and its condition suggests that it dates from the early seventeenth century, which is much later than the Shakubu-ji blade (fig. 133). It is easy to imagine why this tooth configuration was rarely used. Each sawyer would have had to stretch and contract his body and arms to their fullest extent at each stroke to make the blade work efficiently, a very tiring process. As a result, the actual cutting edge was decreased in length, only 1.4 meters in the case of the Kimura Museum blade. Moreover, the apprentice positioned below would have done all the physical work, while the master above would have merely guided the tool along the inked lines on the upper surface of the timber. It is easy to see why the unidirectional frame-saw was less common and efficient than the splayed-tooth type. It is probable that the two-man unidirectional frame-saw was an enlarged version of the smaller, single-person frame-saw common on the Asian mainland but not adopted by the Japanese.

The best evidence of the influence of the frame-saw on architectural form is to be found at the Ryōgen-in, an important subprecinct of Daitoku-ji, the great Zen temple on the northern edge of Kyoto (fig. 134). The main building of the

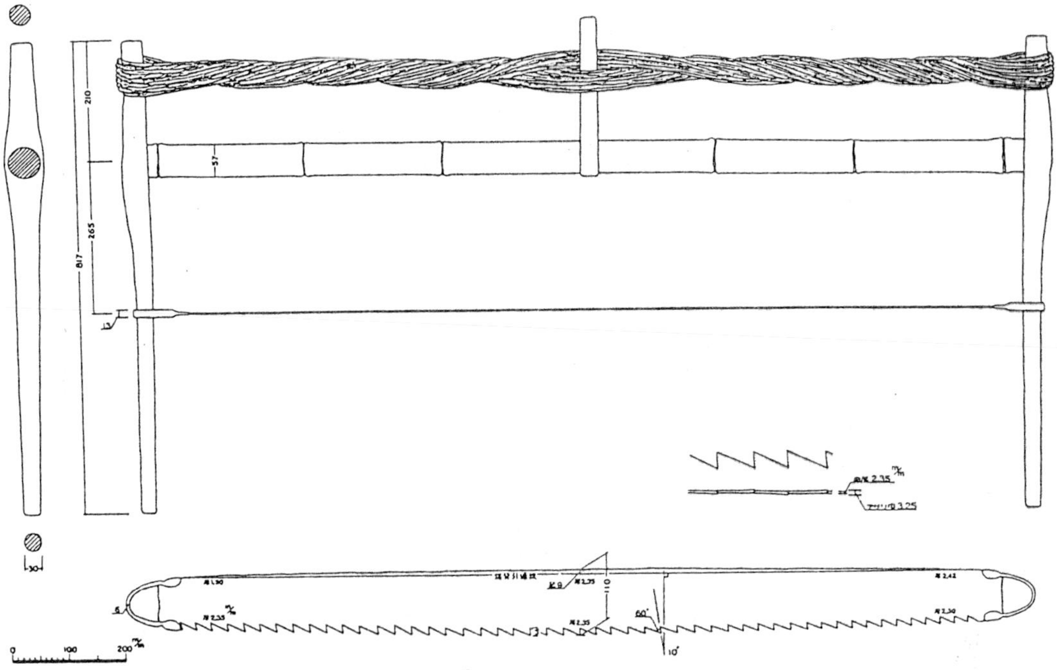

Fig. 133. Two-man frame-saw (*oga*). Early seventeenth century.
Kimura Ethnology Museum, Imari City, Fukuoka Prefecture.

Ryōgen-in, known as the Kyakuden or Hōjō, was constructed around 1517 and is the oldest example of a Zen meditation hall and abbot's quarters still extant. Surrounded by timeless rock and moss gardens representing the spiritual universe of Zen, this simple rectangular structure with its cypress-bark shingle roof, plain wooden floors and austere sumi paintings, still serves as the focus of rigorous retreats and meditation.

The Ryōgen-in Hōjō is one of the few surviving buildings constructed in the 150 years between the introduction of the frame-saw and, in residential architecture, the general use of tatami mats which rendered timber flooring obsolete. In the central chamber, the *shitchū* where prayers and oblations are still offered, the floorboards provide mute testimony to the new freedom bestowed on carpenters by the frame-saw in the fifteenth and sixteenth centuries. The boards, now dark with age and use, are 6 to 7 meters long and more than 50 centimeters wide, smooth on both top and lower surfaces. The widest board, which is proudly placed at the center of the chamber to lead the eye toward the altar (fig. 135), is no less than 71.5 centimeters in width. Even though they are of straight-grained cypress, it would have been impossible to split them to such generous and consistent dimensions. These boards could have been cut only with the frame-saw.

Written records confirm that frame-sawyers were at work on other building

Fig. 134. Exterior approach to Ryōgen-in Subprecinct, Daitoku-ji, Kyoto. Early sixteenth century.

Fig. 135. Detail of floorboards cut with two-man frame-saw for the Main Hall (Hōjō), Ryōgen-in Subprecinct, Daitoku-ji, Kyoto.

projects contemporaneous with the construction of the Ryōgen-in. A ridge plaque from the Takane Jinja, Shizuoka Prefecture, establishes that "oga-hiki" worked on the project under the direction of the chief master builder and completed the shrine in 1519. By the early sixteenth century, therefore, sawyers were unquestionably active at regional construction projects and at important temples in the Kyoto region.

Information about the wages and conditions of employment of sawyers in the sixteenth century may also be found in contemporary documents. In 1568 an official order issued by the Hōjō, the military overlords of the Kanto provinces, to the head of a family of hereditary master sawyers specifies that two oga sawyers were to receive approximately 31.5 kilograms of barley (8 *kan*, 40 *mon*) in return for thirty days' service. Half the barley was to be issued immediately but the balance was to be withheld, presumably pending satisfactory completion of the required work. The order, addressed to the "head of the Fujisawa sawyers" (Fujisawa *oga-hiki-gashira*), establishes that sawyers were organized by this time into hierarchically ordered groups under an official head with a recognized title.

TOOLS OF THE ESTABLISHMENT— CARPENTRY IN THE SERVICE OF THE STATE

Building activity accelerated frenetically in the late sixteenth and early seventeenth centuries as the hundred years of civil war gave way to more stable government under strong warrior leaders who unified the country. Oda Nobunaga, Toyotomi Hideyoshi and Tokugawa Ieyasu, in turn, tightened the reins of national control, culminating in the founding of the Tokugawa shogunate (1603–1867). The extensive resources of the nation were mobilized for the building of a new physical establishment to house and manifest the new political order. Kyoto reemerged from the ravages of the Ōnin War as a glittering imperial metropolis, while Edo rose from the marshes of the Kanto Plain to become the headquarters of the shogunate and the symbol of the new age.

The art of building replaced the art of war as the national industry. The organization of carpenters into trade associations or guilds in the new cities occurred as part of the replacement of the general guilds (*za*) of the feudal domains, with specialist craft guilds or associations (*shoku-nakama*) based on a type of free enterprise. Craftsmen in the building trades comprised about one-third of the entire artisan class; carpentry was by far the largest single component of the building industry. There were normally five times more carpenters than plasterers and tatami-mat makers, and ten times more carpenters than sawyers, stonemasons, and cabinet and screen makers, yet wages in the different building trades were broadly comparable. In 1657, following the Meireki Fire which devastated much of Edo, the government regulated the wages of master artisans in order to control costs. The top level of carpenters was to receive 3 *momme* of silver (11.25 gm.), and a rice allowance, for each day's labor. This amount, which was the same as that specified for master roofers, master masons and tatami-mat makers, could buy about forty liters of rice, which was quite generous. Master sawyers, however, were to receive only 2 *momme* or silver, plus the rice allowance, for their daily labor. Throughout the Edo period, wages did not change significantly. At the rebuilding of Edo Castle in 1844, a carpenter's wage was still 3 *momme* of silver plus food allowances, but there was provision for an additional 2.7 *momme* for overtime and urgent work.

The government took measures to regulate the trades as part of its policy to consolidate national control. In 1616 the period of apprenticeship was fixed by national government law to a term of three years. In 1627 this was extended to ten years, a long apprenticeship, perhaps reflecting more realistically the complexity of carpentry traditions by the seventeenth century. By the end of the century all restrictions on the length of apprenticeship were lifted. An edict of 1698 states that "even service over successive generations" was permissible if in accord with mutual agreement."[6] The population of the castle towns was zoned according to status and profession to reinforce the structure of the political and social order. In the city of Edo, which had over a million inhabitants by the early eighteenth century, carpenters lived south of Nihonbashi toward what is now the Yaesu entrance of Tokyo Station. The sawyers lived in Kobiki-chō, now Ginza 4-chōme in the heart of modern Tokyo. By the end of the period, many of the carpenters lived in the vicinity of present-day Kanda.

The Way of the Carpenter broadened and diversified in response to the strong demand by the new overlords for monumental structures and refined villas. Families of elite master carpenters, like the Heinouchi, the Kōra and the Nakai, became the preeminent architectural firms of the establishment, translating the worldly ambitions of rulers into castle keeps and gilded palaces. They ran complex government building bureaucracies and presided over vast armies of craftsmen. Building projects of this period were extraordinarily labor intensive. The official government records for the reconstruction of the Nikko Tōshō-gū in the mid-1630s show that it required some 4.5 million days of labor to complete this grandiose mausoleum complex.[7] Some three hundred and eighty thousand carpenters were engaged in the project. The Yōmei-mon, the two-story gatehouse guarding the inner precinct and possibly the most vigorously decorated structure in the whole of Japanese history, required 55,970 days of labor (from 1634 to 1636) by carpenters to complete (fig. 136).

The Nikko shrines are not unique in the construction extravagances of the Edo period. More than thirty carpenters were needed per square meter for the rebuilding of the vast audience halls of the inner palace at Edo Castle, including the Great Audience Hall (Ōhiroma) in 1751. Study of the gilt-lacquered coved and coffered ceilings of the audience halls of the Palace of the Second Compound at Nijō Castle, still standing in Kyoto today, reveals the infinitely complex processes of timber preparation, assembly and decoration that were demanded for the pomp and ceremony of state.

The chief master builders, entrusted with such complex and expensive projects, received fixed salaries and formal status in the shogunal court hierarchy. The head of the Nakai family, which supervised all official building projects in Kyoto and the surrounding Kansai district, had an annual income of 1,000 *koku* (180 kl.) of rice, equivalent to that of many of the mid-ranking samurai who served in official capacity for the shogunate. Kōra Munehiro (1574–1646), the Chief Master Builder (*daitōryō*) in charge of the construction of the Nikko Tōshō-gū and the keep of Edo Castle in the 1630s, had a yearly stipend of 100 *koku* and was honored with the bestowal of the title of Lord of the Province of Bungo (present-day Oita Prefecture) for his architectural services to the state. Munehiro was both a master carpenter and a master sculptor, indicative of the stylistic propensities of the age. His enormous sixteen-meter sculpture of a tiger, prowling through a bamboo grove, carved on the gable of one Edo mansion is reputed to have frightened away the fish in nearby Shinagawa Bay.

Fig. 136. Yōmei-mon, Tōshō-gū, Nikko.

New Tools of the Mind—Design Manuals

An important consequence of the national construction projects like Nikko and Edo Castle was the compilation of manuals of building design. Prior to this time, carpenters' workshops left few written records because the process of transmission of their skills had been primarily oral and behavioral. From the later sixteenth century, however, the demand for large-scale buildings of unprecedented complexity exceeded the supply of trained master builders, as can be deduced from the labor figures for the Nikko and Edo Castle projects. The workshops in government service had, therefore, to resort to written manuals to ensure rapid and accurate training of new carpenters and the successful completion of building orders. These manuals, known as hinagata-bon ("pattern books") or hidensho ("secret hereditary writings"), included both technical drawings of buildings and written descriptions of the kiwari proportions to be employed in their design and construction (figs. 27–30).

The carpenters' manuals became the most important tools of the mind in the Way of the Carpenter. They consolidated long-established traditions in response to new political and social circumstances and prompted greater reliance on written documents and technical drawings in building practice.

The most widely known of all the design manuals is *Shōmei* ("Elucidation of craft") compiled in 1608. It represents the point of transference of the closely guarded oral and behavioral traditions of a family of master carpenters into recorded form in response to what its author refers to as "a variety of mistakes in carrying out the building styles of our family in recent years." Its empirical instructions were based on actual practice but they established normative standards for future building design for the next two hundred and fifty years of the Edo period. *Shōmei* was compiled by Heinouchi Masanobu, head of one eminent family of master carpenters working in Kyoto. Comprised of separate handwritten scrolls on gateways, residences and palaces, pagodas, temples and shrine buildings, the manual was based on the teachings received from Heinouchi Yoshimasa, his father and head of the family. Yoshimasa himself added a severe injunction to the colophon in 1610, warning that those to whom its contents were entrusted must maintain the secrecy of the traditions of design it explains.

These records were thus a closely guarded secret within each carpenter's household, but less than fifty years after the compilation of *Shōmei* all this changed dramatically. From the middle of the seventeenth century, carpenters' records were printed using the newly popular woodblock print format and were put into wider circulation among the ordinary carpenters of the towns and cities. This had a devastating effect on the closely guarded traditions of the elite carpenter's families but contributed to the rapid dissemination of their design traditions throughout the populace and regions. It was bad for some carpenters but generally good for most buildings.

The *Buke hinagata*, ("Pattern book of samurai architecture") was the first of these published design manuals (fig. 27). It was printed in woodblock form in 1655, and a second edition followed in 1658. It consists of three separate volumes: the first devoted to warrior residences, including their ceremonial gateways, audience halls, Nō stages and stables; the second to shrine architecture; and the third to interior design. It was compiled by Segawa Masashige, a master carpenter working in the city of Edo on many of the stately mansions of the regional lords. This design manual became the handbook for rebuilding Edo after the disastrous Meireki Fire of 1657 in which as much as 80 percent of the city was consumed, a fact which undoubtly prompted the reprint edition. Trained carpenters were in short supply, and new members of the profession had to be apprenticed rapidly to carry out the herculean task of rebuilding one of the largest cities in the world. This manual established standards for architectural design. It defined the architectural order which, in turn, became the physical definition of the Tokugawa political order.

New Cutting and Finishing Tools

It is no coincidence that during the Momoyama and early Edo periods some of the most remarkable buildings in the history of Japanese architecture were built, including the majestic palace of Nijō-jō, the towering castle of Himeji, the refined Katsura Imperial Villa and that exuberant mausoleum at Nikko. Though disparate in form and purpose, the architectural style of all these buildings was decisively influenced by new cutting and finishing tools. Further innovations in the design of saws revolutionized the methods of turning raw lumber into architectural timbers, while the adoption from China and adaptation of the block plane had a dramatic effect on the finish of wood surfaces. Together the new saws and planes significantly altered the design of buildings and speeded up the construction process. The saws and planes of this era gave carpenters greater freedom for expression in wood surface design; at the same time, it freed them from the worst drudgery involved in the preparation of timbers.

Invention of the One-man Ripsaw

For all its importance in the ripsawing of timbers, the frame-saw introduced from China in the preceding Muromachi period had serious disadvantages. It required two men to operate and its great length made it difficult to forge. In the late sixteenth century, sawyers in Kyoto solved both these problems by inventing the maebiki-oga. This was a short-bladed, easy-to-forge saw which could be operated efficiently by a single sawyer (pl. 10; figs. 137 and 138). This new ripsaw with its distinctive, high crowned blade weighing as much as four or five kilograms drove its large raked teeth aggressively through the timber and soon superceded the two-man frame-saw as the mainstay in lumber preparation.

The development of this saw underlines the Japanese talent for tool invention in response to specific buildings needs, in this case the demand for castles such as Edo, Osaka and Himeji, and some of the largest temple buldings constructed in Japanese history, like the Great Buddha Hall of Hōkō-ji in Kyoto. These magnificent, monumental structures were testimony not only to political ambition and religious faith but also to the efficiency of the new type of ripsaw. The two-man frame-saw was too labor intensive to cope with the escalating demand for timber created by massive national-scale construction projects and it was impossible to forge in sufficient numbers to meet the demand.

The one-man ripsaw took its place at the heart of building industry with the construction of the Great Buddha Hall of Hōkō-ji in Kyoto. This massive structure, completed in 1596, was created as a symbol of the power of the national hegemon, Toyotomi Hideyoshi. *Gushi kenki,* a record book of master carpenters active in the Kansai area, includes a section describing the construction of Hōkō-ji. It notes that "the oga is operated by two persons for sawing. . . . However, at the building of the Great Buddha Hall of Hōkō-ji a [saw] called the maebiki was developed and the tōryō set ten to twenty sawyers to work [with this tool]."[8] This description provides a firm documentary basis for dating the appearance of the new saw at important building projects to the late sixteenth century, the height of the era of spectacle and splendor now known as the Momoyama period.

By 1712, according to the *Wakan sansaizu-e,* the one-man ripsaw commonly measured some 61 by 35 centimeters (2 *shaku* long and 1 *shaku,* 1 *sun* wide). As depicted in the encyclopedia, the blade was thick but straight, flatter in profile than the great high-arched instrument it was later to become (fig. 132). It may well have evolved from a narrow-bladed ripsaw which was stiffened by an iron

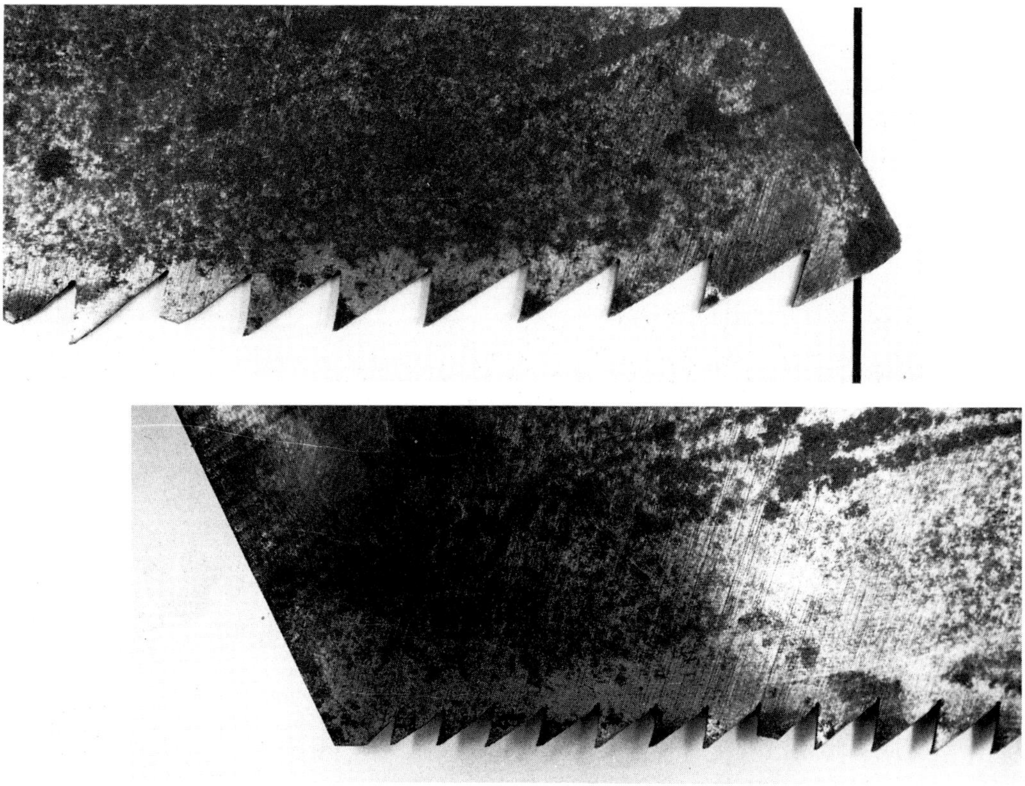

Fig. 137. One-man wide-bladed ripsaw (*maebiki-oga*). Late nineteenth century. *Top*, smaller starting teeth; *bottom*, end finishing teeth.

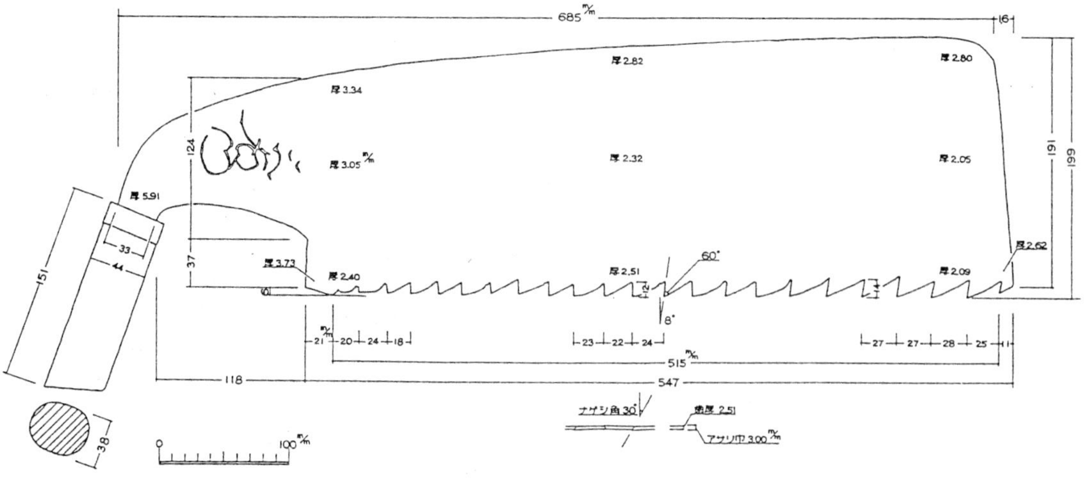

Fig. 138. One-man wide-bladed ripsaw (*maebiki-oga*), signed by Saemon. Early seventeenth-century.

Fig. 139. Comparison of blade width of early seventeenth- and late nineteenth-century wide-bladed ripsaws.

frame similar to that of the modern hacksaw. Experimentation would have proved that a wider blade required no special armature yet readily maintained a vertical position when cutting through thick lumber. As time passed the blades of maebiki became progressively wider. Until special forging techniques were developed, the wider-bladed ripsaws were made from two separate pieces riveted together. By the late nineteenth century, maebiki-oga were almost twice as wide as they were two centuries earlier and had assumed the distinctive hunchback appearance for which they are best known (fig. 139)

According to tests conducted at the Takenaka Carpentry Tools Museum in Kobe, a single person using the maebiki-oga can cut timber as effectively and quickly as two men using the frame-saw. In other words, the one-man ripsaw is twice as efficient as the frame-saw. It is not surprising, therefore, that the maebiki-oga rendered the two-man frame-saw obsolete and quickly replaced the latter at building sites like the Great Buddha Hall of Hōkō-ji. Although the frame-saw is still among the saws included in *Wakan sansaizu- e*, its use by the early eighteenth century was restricted to special tasks such as preparing the thin, wide door-panels of cedar which graced the corridors of shoin architecture

Fig. 140. Sawyers at work in the 1830s using the wide-bladed ripsaw (*maebiki-oga*). "Fuji from the Mountains of Tōtōmi," from Hokusai's print series *Thirty-six Views of Mount Fuji.*

such as the palaces at Nijō Castle. In the harsh, physically demanding world of the sawyer, there was no room for sentimental attachment to outmoded tools.

The sawyer operated the one-man ripsaw either sitting beneath the wood, using the force of gravity to aid the pull cutting stroke, or perched on top of the angled timber. There is a graphic depiction of a team of sawyers using this saw in an 1830s woodblock print from the *Thirty-six Views of Mount Fuji* by Hokusai (fig. 140). One sawyer may be seen balancing on top of the beam, a second sawyer working beneath, while a third man sits to the side patiently sharpening the teeth of a saw with a file. Meanwhile, a fourth figure, perhaps exhausted from the exertion of sawing, gazes at a distant view of Mount Fuji. The wife of one of the sawyers, child on back, offers encouragement.

Rhythm was fundamental to the coordination of movements involved in the cutting process. Although strength was important, a slow, measured action was vital to sustain the effort for long periods and to avoid burning the delicate fiber of softer woods like cedar. The sawyers created their own work songs to accompany their actions and to ensure the correct rhythm in their cutting strokes. One such song, retained in the folk repertoire of Miyazaki Prefecture of Kyushu, is *Hyūga kobiki uta* ("Song of the sawyers of Hyūga Province"). Its words provide insights into the trials and tribulations of the life of the sawyer. It was sung by Hyūga sawyers as they worked and was later disseminated through the provinces of Western Japan by farmers and seasonal workers who traveled in neighboring regions when farmlands were lying fallow in winter. *Hyūga kobiki uta* was sung in free rhythm giving the necessary flexibility for synchronization with the work at hand. The repetitive "*ya-re,*" heard at the beginning and inserted throughout the song as a refrain, served to coordinate the actions of the workers. The main verse of the song translates:

> *Ya-re ,Ya-re !*
> The work of the sawyer is harder than that of the daiku.
> Don't ever become the wife of a kobiki, young maiden!
> Sawing good-tempered wood, he'll deceive you, young maiden!
> And you'll end up his wife.[9]

One can but wonder if this had been the fate of the woman seen in the Hokusai print.

The skill of the sawyer with the maebiki-oga is legendary in Japanese architectural traditions. With this ungainly looking saw he could cut a board or panel as thin as 3.5 millimeters (8 *ri*). In a country where timber had become such a precious commodity, minimizing waste was a critical part of the sawyer's profession. A skilled sawyer prided himself on cutting twelve pieces of evenly dimensioned timber from one single log; a less skilled sawyer could manage only ten pieces. This difference in output meant the difference between providing ceiling panels for an eight-mat and a six-mat tatami room from a log of the same size. To provide panels for the eight-mat room, the less skillful sawyer would have to take a new log from which to cut the extra two panels, effectively doubling the cost of the ceiling.

The one-man ripsaw stands out as one of the major technological and craft innovations in the history of Japanese architecture. It is a perfect illustration of the creation of a specialized tool in response to the specific needs of the building industry. Further, it demonstrates the way in which such a tool has unanticipated and far-reaching effects on the nature of architectural design and the organization of human activity. Acknowledging the maebiki-oga as one of the finest achievements in the marriage of man and materials by tools, the Japanese sent a team of sawyers to the Columbian World Exposition, held at St. Louis in 1893, to demonstrate its operation. A beautifully preserved maebiki-oga in the Morse Collection of the Peabody Museum in Salem, Massachusetts, is thought to have been one of these demonstration ripsaws. As it rests in this museum it still conveys a sense of the purpose for which it was created, its row of neat rivets connecting the cleanly shaped steel teeth to the broad blade. It was recently exhibited in Japan, where few of its contemporaries have survived except as wall decoration for pseudo-farmhouse restaurants in Tokyo.

The Crosscut Lumber Saw

Another type of saw to make its appearance in Japanese building practice during this era was the *daigiri*. It was used for the express purpose of crosscutting felled logs to prescribed lengths and came into general use around the turn of the eighteenth century. It consisted of a long blade, usually 2 *ken* or about 3.5 meters in length, and was closely related in form to the blade of the

Fig. 141. Crosscut lumber saw (*daigiri*).

two-man ripsaw, having similar rings at each end of the blade and a splayed tooth arrangement (fig. 141). Like the earlier oga, it was designed for two-man operation but was operated without a tensioning frame. Instead, round wooden handles were inserted through the end rings to provide a secure, two-handed grip for each sawyer. The teeth varied from raked to more vertical but were set wide (up to 4 millimeters) for maximum blade clearance through the thick trunks. The daigiri depicted in the *Wakan sansaizu-e* has a straight cutting edge, described in the commentary as "one *shaku* wide" (fig. 132) but by the later nineteenth century the blades became as narrow as 10 to 11 centimeters and were convex in shape for increased cutting power.

The daigiri became the main tool used by the *somabito,* men who did the felling and logging on remote mountain sites. It also became the common tool for rough crosscutting and was found in every rural community until a generation ago.

Adaptation of the Chinese Block Plane

The second major tool type to undergo dramatic development in the sixteenth and seveteenth centuries was the plane. The *Wakan sansaizu-e* of 1712 states that "the *tsuki-ganna* [push-plane] was first used more than a hundred years ago." The push-plane was a block plane worked with a pushing action by grasping two handles provided toward the front. It was the standard type in use throughout China and south and west Asia and was introduced to Japan by Chinese merchants trading in the ports of Nagasaki and Kobe at some point in the later sixteenth century (fig. 143). The exact date of its arrival is unclear and unimportant. What is significant is that the needs of building practice in the Momoyama and early Edo period converted it into a simple block plane worked by a pulling action in keeping with indigenous preference. The block plane, minus handles, is shown with other tools casually laid on the veranda outside the shop of a shoji-screen maker in the *Rakuchū rakugai zu* screens of 1615–16, formerly owned by the Funaki family. The transition from the push-plane

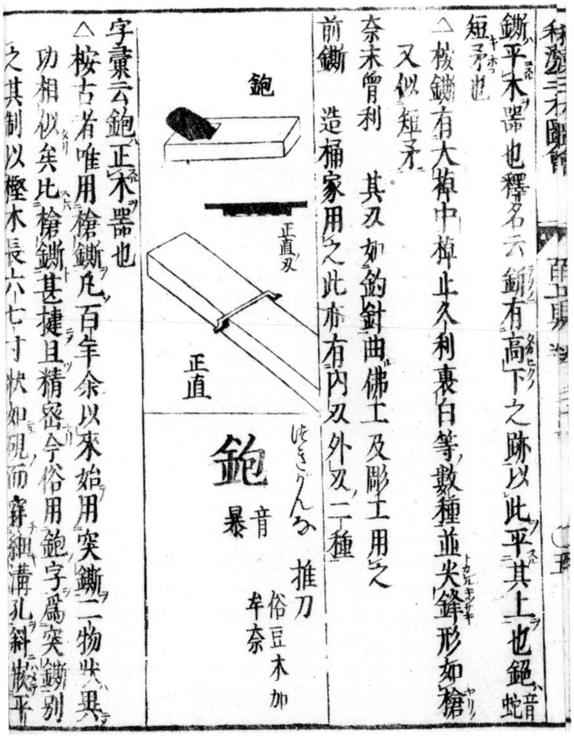

Fig. 142. Section of *Wakan sansaizu-e* (1712) describing block planes. By this date, the plane no longer has handles.

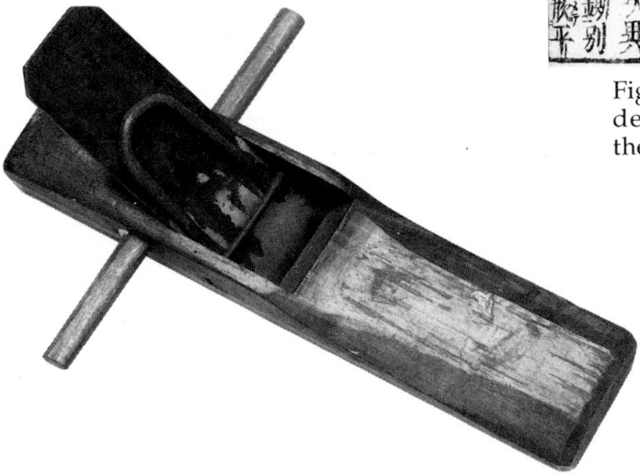

Fig. 143. Modern push plane with handles, Southwest Asia, is similar to the block planes introduced to Japan in the sixteenth century.

with handle is also evident by comparing the *Wakan sansaizu-e*, which still includes the tsuki-ganna as a basic plane type, with the popular encyclopedia *Wakan funayōshi* of 1761, forty-nine years later, which omits all reference to it (fig. 142).

A whole new realm of efficient operation and creative possibilities was opened up by the introduction and assimilation of the block plane into the carpenter's tool set. It rapidly replaced the adze and planing knife for smoothing the surfaces of wood, performing the same operation more quickly and with less effort and—dare it be suggested—with less skill.

The block plane prompted the invention of gouging and grooving planes to make the sliding tracks for shoji and fusuma sliding doors. Hitherto the only traditional method was the painstaking fashioning of the track by attaching battens to the sills. The new gouging planes made it infinitely simpler and faster to fashion the multiple tracks and housings for the screens which give the interior of the Japanese building its special character (fig. 75). Now the way was open for more extensive use of sliding screens and the more flexible arrangement of interior space. At the same time, carpenters displayed their ingenuity by making a variety of molding and beveling planes to create different decorative effects on pillars and other interior detailing.

These developments may seem unremarkable to us now, for the arrangement of interior space using sliding screens has become synonymous with the whole idea of Japanese architecture. However, it was only with the development of the new block planes in the late sixteenth and early seventeenth centuries that the basic method for creating the most common sliding tracks came into building practice. Similarly, the variety of beveling planes, which seem such an indispensable part of the carpenter's tool set, came into use only in the seventeenth and eighteenth century.

Trade Specialization and Tool Diversification

Until the middle of the seventeenth century the most important changes in tools occurred in elite architecture. Thereafter, the focus of interest shifts to the carpenters who built houses and shops for merchants and artisans in the burgeoning castle towns. The style of domestic and commercial architecture became more sophisticated in conscious emulation of the architecture of the elite. Features of the shoin residence and the sukiya "teahouse style," which had been reserved for the palaces and rustic retreats of aristocrats and warriors, became a common sight in the reception rooms of wealthy merchants and artisans, despite frequent government prohibition. The tools and techniques necessary to create the special beveling or consciously rustic blade marks on alcove pillars of the tokonoma alcove and decorative shelving found a place in the building practice of town carpenters as well as sukiya-daiku.

The building profession subdivided along specialist lines to meet the demands of the new urban market. The maker of shoji screens, *byōbu* frames and doors, was known in the seventeenth century as *toshōjishi* ("master of doors and shoji"); by the end of the eighteenth century, he became the more familiar tateguya and set up shop alongside sashimonoshi, carpenters who specialized in making wooden storage boxes with perfectly crafted joinery.[10] The sawyer profession split into separate branches: the fellers, who used axes and the newly developed crosscut saw to harvest timber from the hillsides; and the sawyers, who concentrated their skills on the maebiki-oga to supply squared timbers.

Tools of every kind proliferated as a consequence, reinforcing the separation and identity of the carpentry subprofessions as the techniques needed to master

Fig. 144. Section of *Wakan sansaizu-e* (1712) describing crosscut saws. *Top*, general roughing saws ("boat saws" or *funa-noko*), which are still like the leaf-shaped blades of the medieval period; *bottom left*, carpenter's ripsaw (*gagari*); *bottom right*, coping saw (*hikimawashi*).

them were passed to new generations of apprentices. The *Wakan sansaizu-e* lists convex and concave gouging planes, side-shaving planes for sliding tracks and a dovetail gouging plane as part of the tool set of the carpenter of 1712. Some fifty years later, the *Wakan funayōshi* includes four types of planes for smoothing surfaces ranging from rough to smooth. Chamfer planes are also included as a matter of course.

An additional source of information to the published enclyclopedias such as the *Wakan sansaizu-e* shows that further refinement in saw type occurred during the eighteenth and early nineteenth centuries. This evidence is to be found in later Edo-period paintings like those depictions of artisan tools by Kawahara Keiga, commissioned by the Dutch representatives at Deshima to document Japanese manners and customs in the 1830s. A compact version of the maebiki-oga was developed as a carpenter's ripsaw, the gagari, and was used thereafter for small-scale ripping work at the building site and workshop. It retained the raked teeth of its larger parent but had the convex-shaped back to enhance the effectiveness of the pulling action and to set it apart in appearance from crosscut saws. The crosscut saw itself had undergone some changes in form. Seventeenth-century paintings still show it as having a leaf-shaped blade, but by the time of the *Wakan sansaizu-e* and Kawahara Keiga documentary paintings of tools for the Dutch, it was strictly rectangular in form (fig. 144). Only the general roughing saw retained a more leaflike shape. Other saws to appear at this time were the coping saw, hikimawashi-noko, and the "slender saw" (hoso-noko) used for fine work or cutting holes (fig. 144).

With these developments the carpenter's tool set takes on a more familiar appearance, reflecting the general diffusion of high-culture traditions into the residential architecture of the urban populace in the middle and latter part of the Edo period.

THE MODERN TRANSFORMATION—TRADITION VERSUS THE WEST

The Meiji period was, in many ways, another Nara period: the Japanese leaders set out to modernize state and society on the basis of foreign models, with the same determination and success as they had in the eighth century, and with much the same process of adoption and adaptation. This time, however, the source of inspiration was Europe and America, not China.

The political and intellectual revolution which unseated the Tokugawa shogunal government in 1867 and replaced it with a revival of an emperor-centered one was carried out by junior members of the military aristocracy under the guise of rediscovering Japan's inner historical dynamic—*fukkō*, the "rediscovery of antiquity." But in the next few years those same young men, faced with the responsibility of political and economic management and the establishment of national status in a Western-dominated Asia, quickly became pragmatists of the first order. The slogan fukkō sank out of sight, being replaced by Fukuzawa Yukichi's phrase *bummei kaika*—Civilization and Enlightenment—as a basis for their planned revolution. Article Five of the Charter Oath of June 11, 1868, which reads, "Knowledge shall be sought throughout the world so as to broaden and strengthen the foundations of imperial rule,"[11] became their guiding light. Lavish imitation or modeling on Western lines was not their intention. Okubo Toshimichi, minister of finance writing on behalf of the Meiji government to President Ulysses Grant of the United States at the time of the Iwakura Mission to America in 1871, summed up the theory of government reformation clearly:

> The civilization and institutions of Japan are so different from those of other countries that we cannot expect to reach the declared end at once. It is our purpose to select from the various institutions prevailing among enlightened nations such as are best suited to our present conditions, and adapt them in gradual reforms and improvements. . . .[12]

In 1868 Japan's leaders had set out to modernize rather than Westernize their country. To bring this about, they traveled the world searching for the best models, the most experienced teachers and the most impressive institutions, finding them in the modern post-industrial-revolution countries of the West. These they adapted to suit Japan's special requirements, recruiting specialists from abroad or sending their own young men away for training. It is not surprising, then, that the result reflected the Western models.

Thus the second half of the nineteenth century witnessed the conscious dismantling of the feudal order from above, applying Article Five to every aspect of government and society. Western ideas, systems of government, military organization, culture, industry and technology were tested to meet the new national priorities. By the time of Emperor Meiji's death in 1912, Japan had emerged from her status as a feudal, even backward-looking country, to become the indisputable economic and military power in Asia and a friend and ally of the great British empire in the Anglo-Japanese Alliance of 1902. Domestically, hard work and thrift had enabled the rapidly developing Japanese industries to reinvest more than 15 percent of profits in new plants and still leave a margin to pass on to the people in raised living standards.

New Western-style factory buildings, government and commercial offices, as well as housing and urban development, placed new demands on traditional carpenters and family workshops, requiring speedy adjustment of design, materials and technique. Carpenters who had risen to the top of their profession in the service of samurai lords before 1868 had to turn their hands and their tools overnight to creating a new physical environment suitable for a progressive modern state. Their early efforts emphasized the form rather than the substance of Western-style architecture.

Edo, the old political capital of the Tokugawa shogunate, was renamed Tokyo after the restoration of imperial government, but lived on to serve the new establishment as it had the old. However, the new government needed a new form of physical context. It was impractical as well as a little incongruous to operate the new government from behind the walls of the old feudal mansions. Public buildings in Western styles were essential to the consolidation of power in a consciously Westernizing nation. As part of a broad spectrum of reforms in government, the Meiji leaders set out to rebuild Tokyo along the most advanced principles of contemporary urban and architectural practice of the West. Buildings such as the Tsukiji Hotel, the first official government guesthouse constructed in 1868, were eclectic curiousities (fig. 145). They were only traditional timber-frame structures awkwardly capped with Western-style accretions such as cupolas and towers. They demanded of their builders only imagination, not new tools and techniques. In the 1870s, the Kaichi Elementary School in Matsumoto and the chapel at Dōshisha, a Christian university in Kyoto, employed standard timber-frames held together with conventional Japanese joinery behind their impressive façades of Western weatherboard and brown brick.

The general success of the carefully engineered transformation of Japan into a competitive Western-style nation was physically reflected in the handsome brick and masonry façades of the Tokyo commercial and governmental districts. Many young men were exposed to Western building and engineering techniques by traveling to Europe and America, but many others profited by studying under foreign professors and technical advisers employed on a contract basis by the new Meiji government. Of these advisers, Josiah Conder was the most influential in the field of architecture.

The rebuilding of the Ginza district, after a devastating fire destroyed much of the older downtown sector in 1872 , was the first concerted urban modernization

Fig. 145. Tsukiji Hotel, Tokyo, 1868. *The Far East*, 1872.

project of the Meiji government. The project transformed part of the old Edo artisan quarters into an eloquent symbol of the large-scale commercial enterprises of the Meiji period. It was carried out under Thomas Waters, a British engineer, who was commissioned to build a new Western-style, fireproof district. He laid out a broad avenue southwest from Kyobashi and lined it with neoclassical brick buildings covered with a veneer of portland cement (fig. 146). However, the project produced a classic problem of lack of communication with Japanese carpenters working for a foreign architect. The carpenters could not understand the foreign units of measure nor work with them. It was decided, therefore, to make the English foot equivalent to the Japanese *shaku*. In this way Waters was able to draw up plans using his customary unit of measurement, and the Japanese could carry them out on the basis of theirs. Despite the climatic and seismic shortcomings of building in brick and cement in Japan, the Ginza project had a profound influence on architecture in Meiji Tokyo. It offered the Japanese a tangible model for progress, a physical embodiment of civilization and enlightenment, ushering in an era of brick construction for public and commercial buildings that was the beginning of the demise of timber-frame construction in residential architecture as well.

Rapid advances iron and steel manufacturing in the Meiji period had an impact on architecture, particularly the more utilitarian structures of the new age. Iron-framed buildings, such as the Shimbashi Machinery Hall of the Japan National Railways, dating from the early 1870s, had iron columns proudly proclaiming "Hamilton's Windsor Ironworks Limited, Liverpool" as the origin for their framework. But, by the later 1880s, buildings such as the Shimbashi Factory of the Japan Railway Bureau, completed in 1889 and now preserved at

Fig. 146. The Ginza shortly after construction. *The Far East*, 1874.

Meiji-mura near Nagoya, used locally manufactured cast-iron columns and roof trusses. By the 1890s the Japanese were manufacturing nuts and bolts as well as door hinges for their building requirements. In the first decades of the twentieth century, the Japanese were building massive warehouses of domestically manufactured iron framing and red brick in the port of Yokohama to service the war effort against Russia and their burgeoning international trade (fig. 147). One of the most ingenious applications of Western steel-framing techniques, which also highlights the problems facing traditional construction processes in this era, was at the restoration of the Great Buddha Hall of Tōdai-ji carried out from 1906 to 1913. This great edifice was a sagging, rotting pile by the end of the nineteenth century, despite complete rebuilding carried out under the Tokugawa shogunate from 1688 to 1707 (fig. 148). Neither new timbers nor old building techniques could be employed to save it. Instead, a 19.8 meters wide box truss of imported Shelton steel girders and diagonal braces was inserted beneath the main timber crossbeam of the conventional truss to carry the upper levels of the roof (fig. 149). This Western intrusion was hidden from view by a suspended ceiling of traditional design. The working drawings from the project reveal that the loads and stresses were carefully calculated using newly mastered principles of Western engineering.

The restoration of the Great Buddha Hall demonstrated the limitations of traditional technology by the early twentieth century. One can only be sympathetic with the restorers in resorting to this Western solution. Perhaps this is an apt metaphor for the entire Meiji experience. The scale of the problems they faced was enormous, and attempts to make repairs using traditional means had been unsuccessful. There were no timbers available anywhere in Japan for a

Fig. 147. Brick warehouse, Yokohama, 1911.

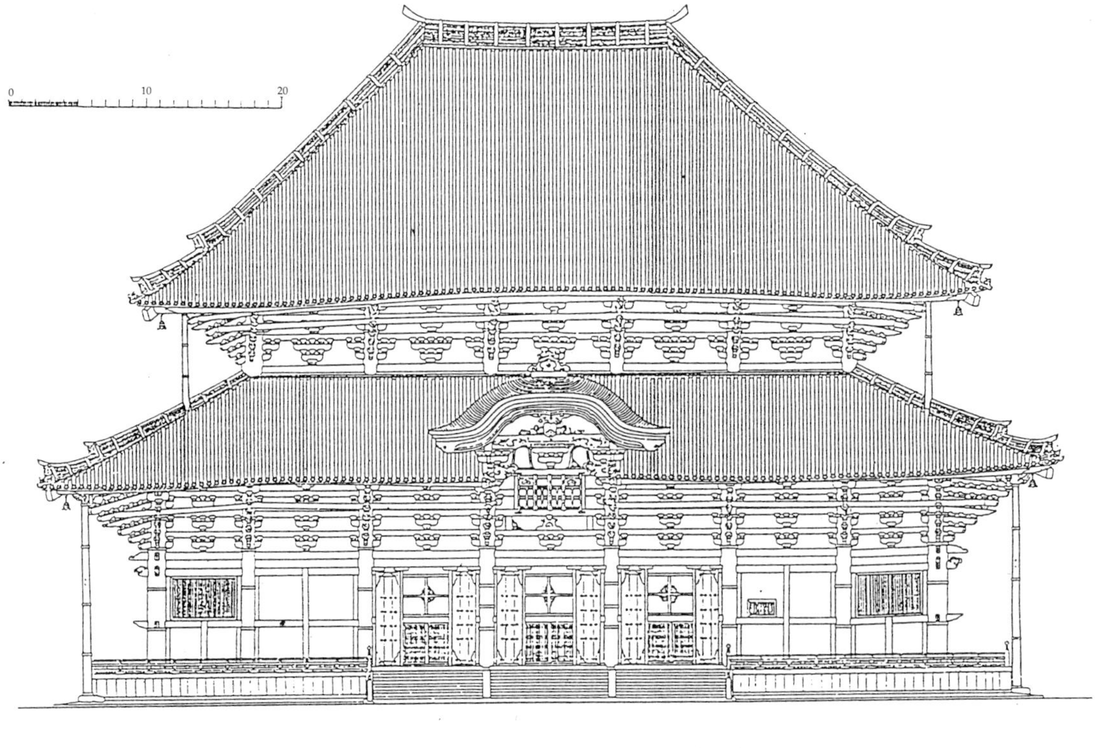

Fig. 148. Great Buddha Hall (Daibutsu-den) of Tōdai-ji prior to Meiji-period rebuilding. *Tōdai-ji Daibutsuden Restoration Report.*

Fig. 149. Great Buddha Hall (Daibutsu-den) of Tōdai-ji. *Top*, transverse section prior to Meiji-period rebuilding; *bottom*, transverse section showing restored roof truss with steel framing. *Tōdai-ji Daibutsuden Restoration Report*.

structure the size of the Great Buddha Hall. Moreover, by the early twentieth century there were no master carpenters capable of building on that scale. There had been no large-scale traditional architectural project in the earlier part of the Meiji period to keep the special skills of monumental construction alive. It had been possible for the Nakai, master carpenters in the service of the Tokugawa, to rebuild the Great Buddha Hall at the turn of the eighteenth century because they were less than two generations removed from the builders of the castles, palaces and religious buildings of the early Edo period. Their fathers and grandfathers would have transmitted to them knowledge of the problems of building the great halls of the temple of Nishi Hongan-ji and Fushimi Castle. The carpenters of the end of Meiji were forced to reach back beyond their collective experience and could not handle this challenge in traditional ways.

Mastery of Western building techniques was most dramatically demonstrated by the grand Versailles-like Akasaka Detached Palace—with its steel frame, imposing masonry façade, sumptuous chambers and central heating—when it was completed in 1909. Intoxication with brick and stone as building materials rapidly turned into a bad hangover when it became apparent that they were not suited to Japan's high humidity and earthquake conditions. Wood continued to be the basic building material in vernacular housing, and the government agencies gradually focused attention upon improving the structural soundness and general durability of conventional timber-frame residences with improvements such as Western-style diagonal bracing.

In the time-honored tradition of the Nara and Edo period governments before it, the Meiji government undertook reform of weights and measures as part of the process of modernization. The Ginza project had exposed the need for effective standards in the building sector. It was an important task of the Meiji government to bring order to the traditional systems of weights and measures in order to facilitate industrialization and trade as well as building practice. After an unsuccessful attempt to standardize traditional weights and measures in 1875, the government determined that traditional Japanese units of measure should be given precise metric equivalents. A decade after the signing of the International Metric Agreement by sixteen nations in 1875, a decision was made to align the Japanese standards with the predominately European-based standard, much to the chagrin of the British and the Americans. It was decided that one meter should comprise 3 *shaku,* 3 *sun.* One *shaku* thus became 30.3 centimeters and 1 *ken,* comprising 6 *shaku,* would be 181.8 centimeters (182 centimeters to the next highest decimal place). These standards were formally adopted in 1885 and promulgated by imperial decree in 1886.

As part of the broad spectrum of changes in the later nineteenth century, the Japanese also set out to establish a Western-style architectural profession. This had far-reaching consequences for the traditional carpenter and his tools. The distinction between design and construction has been a fundamental doctrine of European architecture since the Renaissance. In Europe before the Renaissance, as with Japan before Meiji, there was no separate profession of "architect." Although sophisticated techniques of architectural draftsmanship had evolved in Japan by the Edo period, the building designer remained "on-site" and "in touch" with the techniques and materials of his buildings. The art and craft of building were one indivisible process, as shown in the earlier chapters.

In the Meiji period the first Japanese trained to be architects in the Western style emerged from the tutelage of Josiah Conder and the fledgling Department of Architecture of what later became the University of Tokyo (Kōbu Daigakkō). Influential architect-scholars of this new generation, particulary Itō Chūta, argued strongly along Western lines for the independence of the architect as a design profession separated from the building site and working practice. This was the beginning of the demise of the prestige and responsibility of the master carpenter. Today there is serious reevaluation of this philosophy. Now it seems incontrovertible that the building designer should understand construction materials and proceses, but in the Meiji period the Renaissance notions of the architect began to drive a wedge between carpenters and his traditional design prerogatives.

How were carpenters and their tools faring while all these changes were going on? How were they dealing with the challenge posed by fancy Western-style buildings and by the arrival of the much vaunted Western-style architect?

The survey of carpenters' tools conducted in 1943–44 by the Institute of Labor Science, it will be recalled, revealed that the tools and the techniques of Japanese carpenters were still basically "traditional" as late as the middle of the twentieth century. Augers, brace and bits, spanners and adjustable wrenches were added to the tool repertoire for the special purpose of inserting bolts into framing members. In their daily work, however, carpenters were still using their pull-action saws and planes, their butt chisels with special shoulder bevels for detailed finishing work on lintels and frieze bands, and the square with its scales for calculating curvature, which were at the heart of traditional building practice. Today a daiku may still use a traditional chisel to make Western-style weatherboarding (fig. 150).

Nevertheless, from the mid-nineteenth century important changes had been made to the design of some tools. The last years of the period of Tokugawa government witnessed cross-fertilization of Japanese and foreign technology to create the *dōtsuki*.[13] This saw, which was to become indispensable for making the delicate framing of shoji, was invented in the early 1860s by the simple expedient of taking the standard Japanese crosscut saw and adding a spine in the manner of the Western tenon saw. The result was a tool which proved superior to both its progenitors, combining the advantages of the Japanese pull-stroke teeth with the precision of the reinforced, Western tenon saw.

In the early years of the Meiji era, carpenters, ever quick to realize ways of increasing the efficiency of their tools, added Western-style wing nuts and screws to adjust the fences of plough planes, a decided improvement in strength and durability from the traditional bamboo wedge they replaced.

The double-bladed plane, now the standard type in Japan, was also developed in this era of adaptation. It came into use about the time of the Russo-Japanese War of 1904–5 by the addition of a Western-style capping iron ("chip breaker") to the principal cutting blade (fig. 151). This forced the shavings away from the cutting edge of the blade, reducing torn grain and making the plane easier to handle for the less-skilled craftsman, an interesting comment on declining standards of skill by the early twentieth century.

Fig. 150. A Japanese carpenter making Western-style weatherboarding with traditional tools.

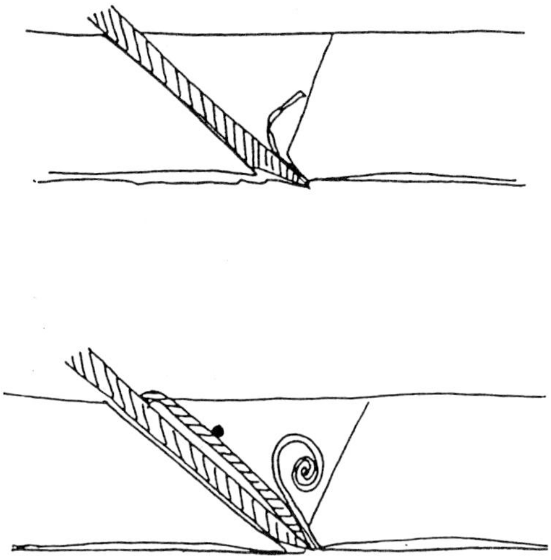

Fig. 151. Comparison of the cutting action of single-bladed and double-bladed planes.

There was also one significant improvement to existing traditional technology achieved independently from Western influence in the late nineteenth century. The double-edged saw, or ryōba, combining rip and crosscut sawteeth, was developed for the First Japanese Exposition for the Promotion of Industry held in 1877. This saw offered a considerable saving in time at the worksite and effort at the forge and came into general use about 1895—a late date in the history of Japanese architecture for one of the tool types now most commonly and erroneously associated with Japanese traditional architecture.

There were important changes in the methods of making tools which contributed to the changing character of the building process. Some were beneficial, most were decidedly not. One of the beneficial results was in toolmaking. As part of the dismantling of the feudal system, samurai swords were prohibited after 1876. Many swordsmiths, faced with unemployment, turned their art to making knives as well as chisel, saw and plane blades. This created a supply of cutting edge tools of unprecedented quality (pl. 9). Never had the Japanese carpenter enjoyed such superlative steel, and never before had the need for this steel been so limited. Toolmaking using imported steel began about 1879, creating plentiful supply but decreasing quality. Domestic production of Western-style nails began in 1898, and machine manufacturing of saws began in 1920.

The Sōgakudō and Meiji Carpenters' Tools—Japanese Technology and Western-Style Buildings

One building project of this era of transformation helps bring the changes to design and carpenters' tools into sharper focus. This was the construction in 1889–90 of the Sōgakudō, which now stands in Ueno Park in Tokyo close to its original location. The Sōgakudō served as the school where most Western-style musicians of Japan were trained until well into the twentieth century and contained the first Western-style concert hall in Japan. It is a rewarding case study because many of the details of its construction were revealed during restoration from 1983 and 1987.

The general impression created by the Sōgakudō is of a competent Western building, so much so that after you have been inside for a short time you wonder if you are in Japan at all. It consists of a central two-story structure with symmetrical flanking wings (figs. 152 and 153), each originally 30.9 meters in length. It is Classical Revival in style, with carefully crafted timber moldings, cornices, small triglyphs under double-hung windows, applied pillasters and a Palladian portico at the main entrance. There is an impressive baroque pediment set into the eave line over the second floor at the center (fig. 154). The interior of the building has a symmetrical arrangement of rooms set along corridors traversing its east-west axis and a concert hall on the second floor which seats 338 people in tiered rows. This auditorium is light and airy with a high, barrel-vaulted inset ceiling braced by tensioning rods (fig. 155). The main roof of the building has rigid triangular trussing and diagonal struts in the best Western traditions.

How did such a building come to be constructed, just two decades after the Meiji Restoration, and who were the men who made it? Of course, it was not an isolated instance of Western-style construction. By the 1880s and 1890s most public buildings were being made in this style either in timber or brick. The building was designed under the supervision of Yamaguchi Hanroku, chief architect to the Ministry of Education. In view of the structural devices employed and the emphasis on Classical Revival forms, it comes as no surprise that Yamaguchi was trained in Paris at the École Polytechnique, the engineering institute established by Napoleon. In 1876, at the age of eighteen, Yamaguchi was selected to be part of the second group of Education Ministry-sponsored students to study abroad. He completed the three-year course in civil engineering at the École Polytechnique, followed by two years' practical experience in Paris, before returning to Japan in 1881. Together with Kuru Masamichi, an early graduate of the Faculty of Engineering of the University of Tokyo, Yamaguchi was to design many of the buildings required by the rapid expansion of Western-style education in Japan in the later 1880s and 1890s. Reform of elementary school education alone needed some 28,000 primary school buildings nationwide. They created the style of educational architecture in Japan. But who created its substance?

The Sōgakudō appears to be entirely Western, but appearances can be deceptive. Three carpenters' tools from the original construction project of 1889–90 were discovered in the ceilings and walls during the 1980s restoration.

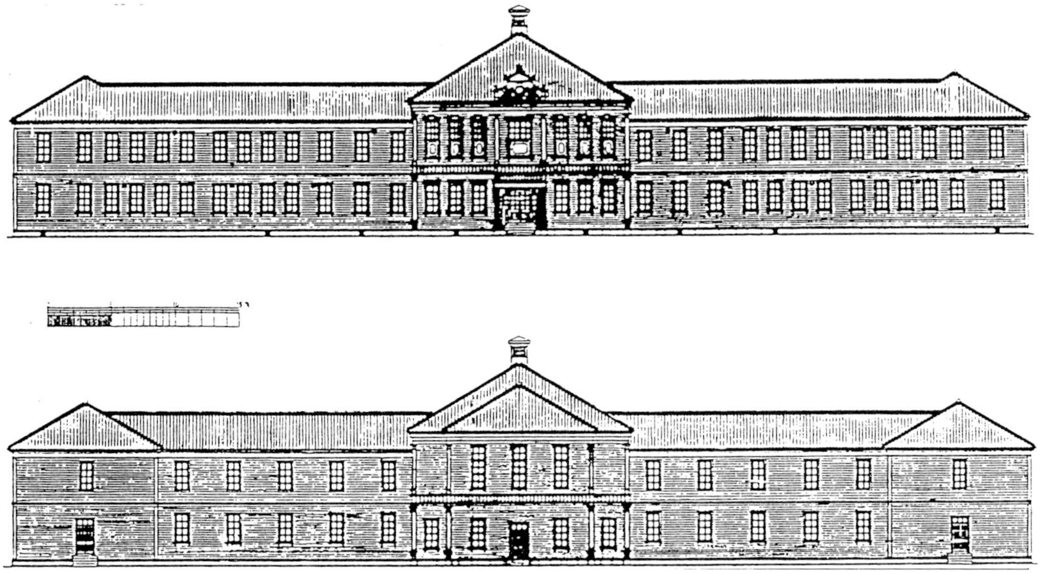

Fig. 152. Elevation of the Sōgakudō, Tokyo, at the time of completion in 1890. *The Sōgakudō Restoration Report.*

Fig. 153. The Sōgakudō after restoration.

Fig. 154. Detail of front façade pediment of the Sōgakudō, 1890. *Sōgakudō Restoration Report.*

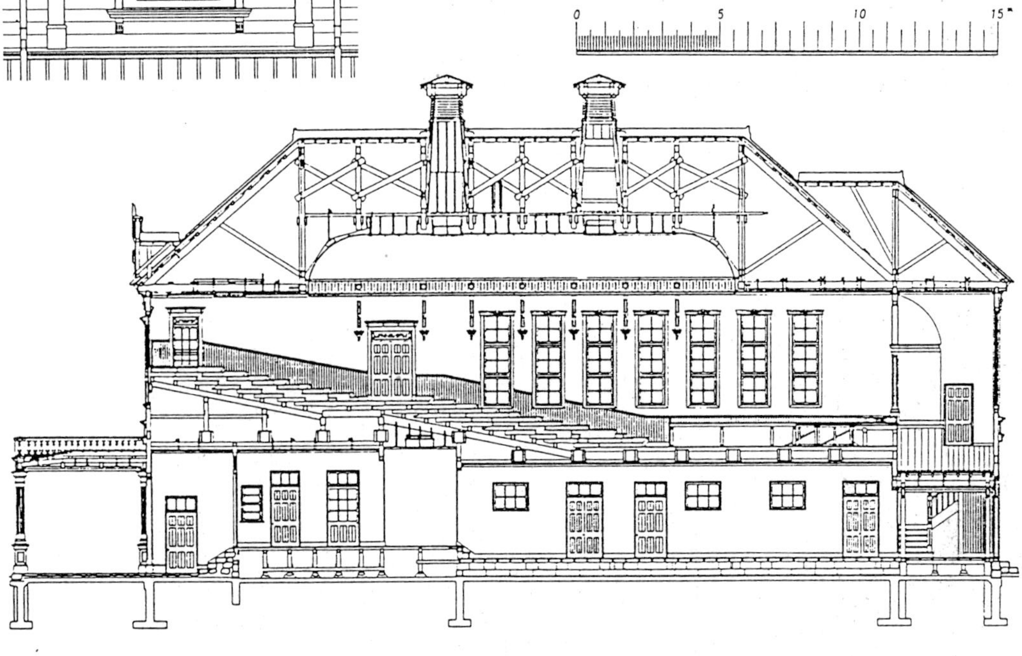

Fig. 155 The Sōgakudō, Ueno, Tokyo,1890. Transverse section showing concert hall on second floor and Western roof truss and chimneys. *Sōgakudō Restoration Report.*

Can we expect evidence of the arrival of the early Stanley plane or the Western push saw in Japan? All three tools are conventional Japanese tools: a plumb-line (sagefuri), a striking chisel (tataki-nomi) with a 1.76-centimeter head and, that most ancient of tools, an adze (chōna) with a 10-centimeter wide blade. Examination of joists and sills of the main staircase of the central hall established that these were smoothed by adze not by plane, an important indication of the application of traditional building techniques to the realization of the Western design of the building, twenty years after the Meiji Restoration was fully launched.

When the building was dismantled for restoration, it was discovered that the timber-frame was held together not by the imported bolts and plates, very much in evidence, but by conventional joinery techniques not so obvious to the eye (fig. 156). The bolts and other metal fittings served merely as structural reinforcement for traditional joints. The strengthening of traditional building frames using Western-style metal fittings became standard practice during the Meiji period.

The restoration also discovered the use of sumi numbering according to the bansuke system on the original timber-framing members, in much the way that bansuke have been found on the remains of pillars of ancient Nara recovered from under the rice paddies. Thus traditional prefabrication, together with labeling, marking and assembly using the plan board, were the basic building procedures employed at the Sōgakudō. Furthermore, during restoration, part of a kensao, the master carpenter's measuring rod, was recovered. As we know, this hand-marked ruler was used by the chief master carpenter, tōryō, to translate the design of the building from the plan board to the building by providing actual 1:1 vertical dimensions. The ruler was usually 6 or 10 *shaku* in length, or slightly over 3 meters, but in the case of the Sōgakudō only a section, approximately 60 centimeters long, reused as a batten in the wall or roof frame survives from this master measure. This kensao shows that the *shaku*, employed by the master carpenters to build the Sōgakudō, was precisely 30.3 centimeters in length. This was the standard Meiji unit set in 1886 as part of the rationalization of weights and measurements. The discovery of a section of the kensao conforming to the Meiji standard suggests that the standardization of measures in Japan was being followed four years after it was promulgated by law, a very different situation from the attempts to standardize measurements in the Nara and Edo periods.

Finally, the enthusiastic, indeed excessive, beveled detailing of door frames and lintels and the execution of moldings and cornices may be Western in form but are of superlative quality as craft, reflecting a delight in detailing which draws from within Japanese carpentry traditions while acknowledging Western influence.

Thus there is compelling evidence of Japanese techniques in the design and the construction of the Sōgakudō. Despite its grand Napoleonic and Renaissance design inspirations by Yamaguchi and Kuru, this most Western looking of Meiji buildings masks fundamental Japanese building methods. The Sōgakudō reveals one of the hidden currents coursing through the veins of Japanese architecture of the later nineteenth and early twentieth centuries. There was no special category of "Western carpenter" brought into existence in the Meiji period to make Western-style buildings. Traditional Japanese tools only were used in the creation of many aspects of the new Western building styles. The means by which Meiji architecture was created diverged less from the well-trodden paths of past building practices—the type of "evil customs of the past" singled out for criticism in The Charter Oath of 1868—than their Western appearance might suggest.

Although the strict apprenticeship system for carpenters went into decline in

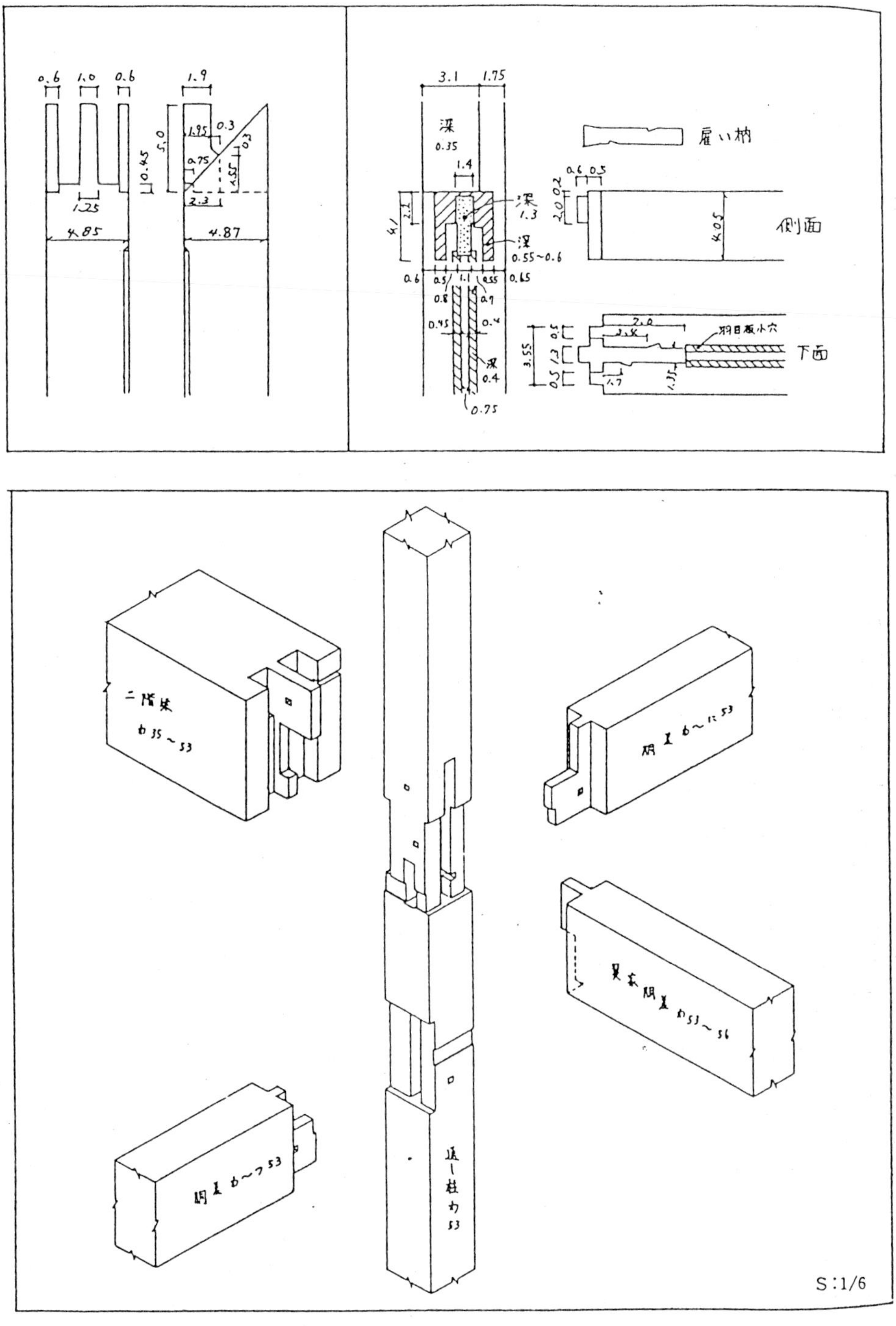

Fig. 156. Sōgakudō, 1890. Traditional Japanese joinery was used to connect the timber frame. *Sōgakudō Restoration Report.*

Meiji with the new social system, and educational reforms such as vocational training as part of the school system undercut the traditional pattern of learning trades, until the turn of the twentieth century, construction projects were still largely in the hands of master carpenters who had been trained under the old apprenticeship system at the end of the Edo period. Although some carpenters were exposed to Western building styles in the treaty ports in the 1850s and 1860s, the effectiveness of this exposure appears doubtful. Most carpenters engaged in Western-style construction in the Meiji period rarely, if ever, met a Western carpenter or bricklayer. Even the carpenters trained in early Meiji by the last Tokugawa generation, in whose hands rested the responsibility of building such "modern" architecture as Tatsuno Kingō's Tokyo Station and Frank Lloyd Wright's Imperial Hotel, retained most of the attitudes and techniques of their immediate predecessors in tradition.

Thus the basic thought processes of Edo period building practice and the techniques of tool use were intact for an entire generation after 1868. One may also speculate on the conceptual congruences between Japanese architectural practices and the forms adopted from the West in the nineteenth century. A master carpenter of the old order was rigorously trained in kiwari proportions. The classical tradition of Europe is based on a similar modular order. It undoubtedly required less mental gymnastics on the part of a carpenter well versed in the ratios of temple design to build the pillars and entablature of the Sōgukudō than might be supposed. Indeed, the fondness for entablatures exhibited by buildings created around 1890, including the Japan Red Cross Building and the Tokyo Postal Communications Office, may relate to the traditional fondness for the cusped gable, kara-hafu. From the later sixteenth century in Japan, the kara-hafu had festooned eave lines to give frontal emphasis to castle and bathhouse alike. It is intriguing that a butcher shop built in Kobe in the 1880s, and preserved at Meiji-mura today, has a perfectly executed neoclassical façade, complete with columns and pilasters, and a conventional Japanese roof gable set above the doorway at the center.

When it came to learning Western techniques of building, the Japanese carpenters resorted to their traditional practice of compiling instruction manuals. In the 1880s and 1890s there was a proliferation of woodblock printed *shin-hinagata* ("new hinagata") explaining Western systems of rigid triangulated roof trusses and the methods for inserting iron bolts into wall frames. As early as 1871 a manual in the traditional style was issued describing the techniques of brick construction, including methods of laying bricks as well as techniques for framing the structures (fig. 157). This manual may have assisted in the construction of the Ginza project under the direction of Waters.

Why was there this obstinate preference for traditional ways of doing things? Was it sheer stubborness, the inherent conservatism of customary tradition, or was there some inherent justification for continuing old methods to build a new age? Viewed objectively, there were many aspects of traditional Japanese building practice which were well in advance of those of the West in the latter nineteenth century. With the exception of large-scale engineering, such as that at Tōdai-ji or for making bridges, Japanese tools and techniques of building still had a compelling logic in the Japanese context. Europe and America may have

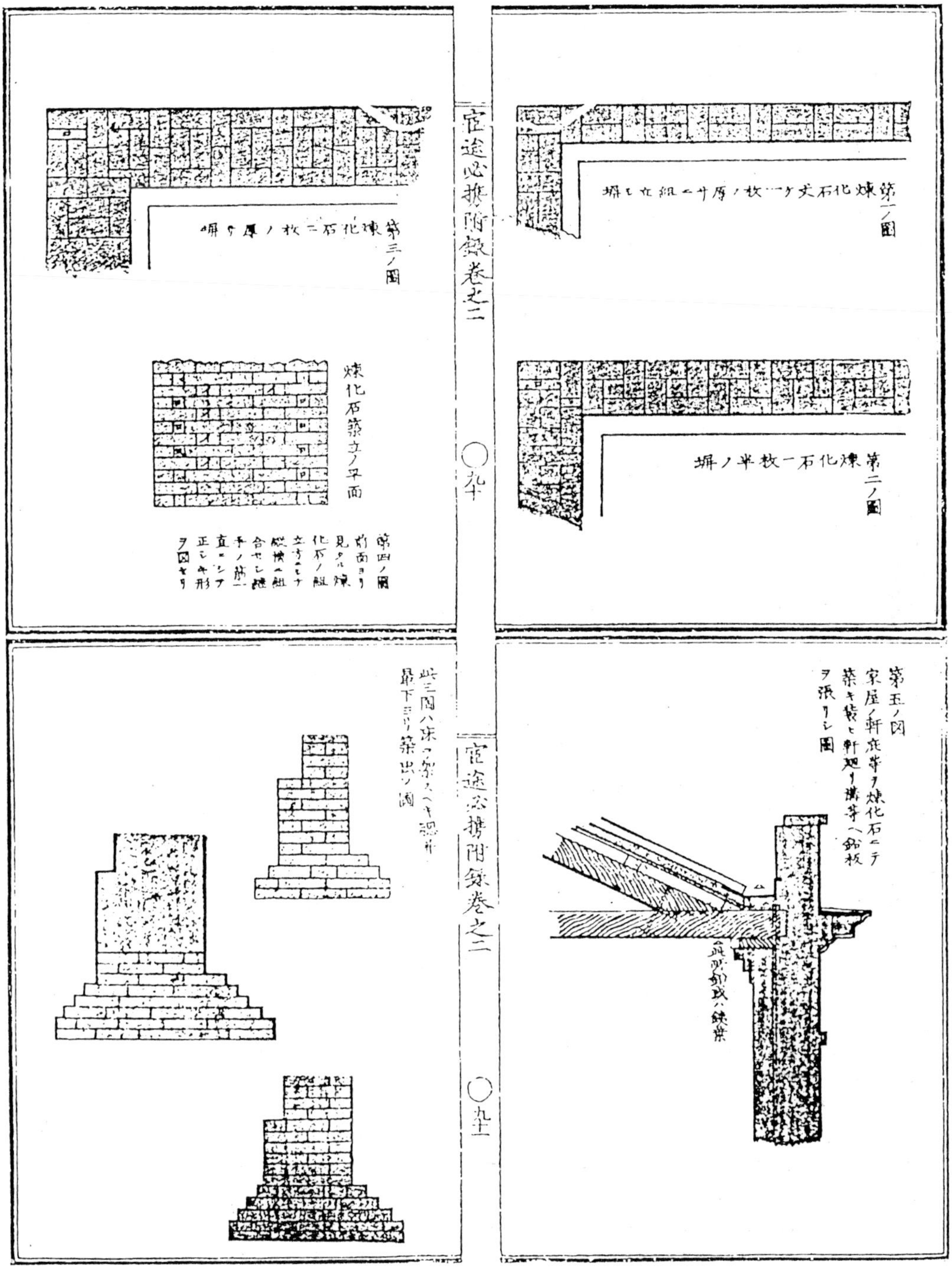

Fig. 157. An 1872 carpenter's manual that is used to explain methods of Western brick construction.

devised new methods of construction using ferrous metals. The West may have been the source of new designs and stylistic inspiration, the point of origin for new ways of life and government which required new containers to house them, but traditional architecture in Japan offered distinctive features, many of which have become ideals for even the most advanced architects working in the West today. Among these are modular coordination of design, a high degree of prefabrication of the parts, and superlative standards of finish.

In the nineteenth and for much of the twentieth century, the Japanese were not as thorough in adopting Western carpentry techniques and framing procedures, and types and methods of tool use, for the simple reason that their own traditional technology was better. They soon discovered the shortcomings of Western-style brick construction in both seismic and climatic terms, as noted earlier in relation to the Ginza project. Moreover, in timber-frame buildings Western-style diagonal bracing proved less than effective. Traditional building design in Japan was based on the "flexible frame" principle now widely used for high-rise buildings. Diagonal bracing makes large flexible frame structures rigid and hence liable to damage in a severe earthquake, as was discovered in the 1923 Great Kanto Earthquake. Western-style diagonal braces and tensioning devices, added to the main shoin complex of Katsura Imperial Villa in the Meiji period, were systematically removed in the recently completed restoration project. They were found to have impeded the earthquake resistance of its original traditional flexible frame structure. Why should the carpenters of the Sōgakudō have abandoned their razor-sharp tataki-nomi for Western chisels? The answer is simple. The laminated steel produced by the master metalsmiths of Japan was superior in every way to anything produced in any American or European factory. The Western push saw, with its thick blade and cumbersome cutting action, offered no advantages to the Japanese already well served by their flexible, fine-bladed pull saws. It is not surprising, then, that today in the West there are master carpenters and builders concerned with quality and craftsmenship who are adopting Japanese chisels and saws for their most refined work.

HAND TOOLS VERSUS MACHINES IN THE POSTWAR ERA

"In the olden days, laborers possessed their own hand tools to engage in production," wrote Abe Isoō in 1901. "They lived an independent life. However, today intricate machines are invented, and unless one can invest a substantial sum. . . . we cannot readily obtain such machines. In this way, one by one, workers are forced to leave their handicraft works at home and engage in factory work."[14] Such was the situation in the first decade of the twentieth century when Japan had accomplished most of the material and political goals set by those responsible for the implementation of the Meiji Restoration of 1868. Abe's words were to prove even more apt in the postwar era, for much of Japan's success as an economic giant has been obtained at great cost to "handicraft works."

From the early nineteenth century, the replacement of hand tools and

Fig. 158. Tokyo today. A way of life alien to the Way of the Carpenter.

techniques with power tools and mechanized mass production proceeded apace in Europe and America, but in Japan the destruction of craft traditions has occurred more recently and rapidly, primarily since the early 1940s. The complex, interrelated building professions were still intact until the time of the war in China and the Pacific (1937–45). The 1943–44 survey of the Institute for Labor Science found that carpenters still used a considerable number and variety of traditional hand tools in their professional work. However, an entire generation of carpenters, not to mention roof tilers, plasterers, stonemasons and others, was swallowed up during the eight long years of the war, as were stocks of building materials and Japanese cities themselves. For the defeated nation there was a loss in confidence in tradition. Pragmatic Westernization, largely in the form of "Americanization," became the order of the day in the decades after the end of the conflict. Industrialized mass production of housing components was the only solution to the chronic shortages of dwellings, skilled labor and materials. Precast concrete panels, plastic and metal sheathing, and power tools together created a new prefabricated vernacular to rehouse much of the urban population.

In the space of a single generation, traditions built up over many generations were swept away. There was little room for creative compromise between new and old in the pressing circumstances of postwar reconstruction (fig. 158). Throughout the 1960s, the priority given to rapid economic growth not only caused environmental havoc; it also created a cultural climate in which many craftsmen, from roof thatchers to master carpenters, vanished and their skills died with them. They were unable to compete with the machine-made products of the modern economy or to convince their children, the "new breed," or *shinjinrui* of the postwar generation, that there was dignity and fulfilment to be found in the traditional professions. The Way of the Carpenter was diverging from the way of most Japanese. It had suddenly gone from being a pillar of society, central to its industry and culture, to become a curiosity of another age. Cost-efficiency was part of the problem, but the loss of respect and self-esteem through deprivation of design initiative was its root cause. The separation of the designer-architect from the carpenter-builder meant that the organic nature of the traditional architectural process was destroyed. Carpenters were demoted to the role of manual workers in a society which sets great store on high educational attainment.

Most building craftsmen now begin their training after graduating from high school, because compulsory education to the age of sixteen as instituted under the Allied Occupation virtually eliminated the traditional system of apprenticeship commencing from childhood. The modern carpenter may be better educated in general than his predecessors but his manual skills are regarded as "retarded" by the carpenters trained from early adolescence, along with a dozen or more fiercely competitive peers, under the traditional system. Today the fully trained apprenctice is already twenty-three or twenty-four before he can begin to establish himself in his profession. The trade schools *(senmon gakkō)* for carpentry and construction emphasize the mastery of power tools and of two-by-four and panel construction, not traditional carpentry techniques. The master metalsmiths' forges, which produced the finest laminated steel the world has ever seen, are now cold while the master technologists of Japan today concentrate their efforts on producing high-quality Ryobi radial arm routers and Hitachi band saws. Traditional building techniques are reserved for temple and shrine construction and creating Japanese-style restaurants, *ryōkan* (inns) and the villas of the wealthy. Even at these sites power tools are in evidence. Today, as the second postwar generation of carpenters grows to maturity, the typical carpenter rarely uses more than thirty of the traditional hand tools. Senior carpenters may still own several hundred tools, but most lie unused in the toolbox. Carpenters now use sashigane marked in centimeters and millimeters, and many younger carpenters do not recognize *shaku* and *sun*. Those precious distillations of Heian- and Kamakura-period design experience, the scales on the rear of the main arm of the square, are today rarely used to calculate curvature or lumber dimensions. Such mainstays of the Way of the Carpenter as the adze have long since disappeared from the modern construction site.

The widespread dissemination of power tools, first manufactured in Japan in 1957, has not only changed work practices; it has altered attitudes as well. Power tools have fractured the long-established and creative relationship between master carpenters and their hand tools. As one old carpenter remarked recently, "The children of the neighborhood used to come and watch us as we worked. When we started using power tools we had to shout *abunai* (watch out!) and warn them to go away." The happy scene of children playing around the

Fig. 159. Japanese "industrialized" houses are up to 80 percent factory prefabricated.

toolbox at the building site, captured in that memorable early fourteenth-century Kasuga scroll (pl. 2), recedes even further into the past.

The Age of Industrialized Housing has come. Japan now boasts the most sophisticated housing industry in the world, its industrialized techniques producing room-sized units with the same masterly precision as automobile units rolling off factory assembly lines. As much as 15 percent of all houses built each year are made in factories. Whole bedrooms, kitchen units, bathrooms complete with electrically warmed toilet seats, traditional Japanese tatami rooms furnished with tokonoma, and entrances complete with ornamental chandeliers follow each other off automated production lines. Once foundations and plumbing are prepared, these factory prefabricated units can be assembled at the building site in less than a day (fig. 159). Despite their mass-produced form, these houses allow for individual variation, for individuality despite industrialization. CAD-CAM computerized graphics with automatic cost-estimation and inventory control allow each client to adjust basic designs to his or her own needs. The exteriors are attractive and finished with the quality expected from Japanese industrial products. The interiors are indistinguishable from those of an architect-designed house and are the ultimate in machine-made housing, completely attuned to the needs of the Japanese "salaryman" and his family. Moreover, they offer higher levels of comfort and safety against earthquake, wind and fire for a lower cost than does conventional timber-frame and concrete housing.

Industrialization of building components has had a significant impact on conventional housing. Contemporary houses are frequently designed around the dimensions of standardized manufactured components. In traditional practice, the fittings carpenters measured the completed framework of the building and then made the necessary sliding screens, frames and furnishings. Now the custom builder is taking the dimensions of manufactured aluminum window frames, prefabricated bathroom units, and system kitchens and designing the building around them.

Like the way of life of which it was a part, the Japanese timber-frame residence is becoming a thing of the past, even though surveys show that 78 percent of Japanese still would like to live in a traditional timber-frame house. There has been a startling decline in the construction of Japanese-style, timber-frame buildings and a rapid increase in the industrialized vernacular and steel and concrete high-density housing. In 1963, wooden structures comprised 86.2 percent of housing units constructed nationwide. By 1983, just twenty years later, this had dropped to 46.3 percent. This figure included the popular Western-style houses made from imported two-by-four studs, making the actual number of conventional Japanese houses even lower than the percentage suggests. In Tokyo, Japan's biggest housing market, the 1987 figure for conventional timber-frame residences was only 27.8 percent; the soaring cost of land is forcing the Japanese salaryman away from a free-standing home to a concrete condominium unit.

Traditional carpentry techniques, therefore, are in danger of becoming extinct as Japan turns away from its wood-building traditions to embrace new materials and methods of construction. In the shrinking private house market, the small-scale local firms, the *daiku-kōmuten* that comprise more than half of Japan's construction companies, are going out of business, particularly in the large cities. The houses they build are unable to compete in a market dominated by large, diversified and imaginative corporations offering well-marketed industrialized houses and attractive package deals of house, land and loans. Traditional joinery techniques require training and time to execute, increasing the cost of housing above market rates now set by the machine-led sector.

The design and hand tool techniques which were a universal part of traditional building practice are now confined to a few kōmuten or design-build firms which specialize in the preservation of historic buildings or in making the highly refined sukiya architecture. There are several associations of traditional craftsmen, architects and academics which act as umbrella organizations, seeking ways to revitalize the Way of the Carpenter in the modern context through workshops and training, conferences and publications, educational reform and fund raising. These include the Kyoto Guild of Master Builders (Kyōto Dentō Kenchiku Gijutsu Kyōkai), based in the old capital, the Japanese Association for Conservation of Architectural Monuments (Bunkazai Kenzōbutsu Hozon Gijutsu Kyōkai), based in Tokyo, and the Timber-frame Architecture Forum (Mokuzō Foramu), at the University of Tokyo.

In these desperate circumstances, an odd marriage of convenience has taken place between some small-scale local builders and machine tool manufacturers. The progeny of this marriage has been computer-controlled joinery machines

Fig. 160. Use of joinery machines at Kinoshita Kōmuten, Saitama Prefecture.

(fig. 160) whose advent marks the latest chapter in the history of carpentry tools in Japan. These machines were developed in the late 1970s by ingenious local carpenters working in collaboration with machine tool manufacturers such as Miyagawa Kōki and Makita Company, the Japanese equivalent of Black & Decker Corporation in the United States. This new generation of tools makes the special spliced and angled joints which lie at the very heart of the Japanese building tradition, producing a quick-fix solution to the crisis of cost-effectiveness in the conventional housing sector. A technician with no training as a carpenter is able to use these machines to make the complicated gooseneck mortise-and-tenon joint for splicing sills, taking only twenty seconds complete a joint that requires twenty minutes of precise, arduous work by a trained carpenter using hand tools. The machines also eliminate need for the comprehensive measuring and marking of each timber which was such an important part of traditional carpentry practice. It is no longer necessary to mark each pillar and beam to show the location and configuration of each joint. Now the carpenter simply provides a centering mark for the machine operator and stamps the wood with a symbol indicating which type of joint is to be used, making it possible to prepare about forty pillars a day. However, it is interesting to note that a plan board is still drawn up from the detailed architectural drawings, and traditional marking tools, including the sumitsubo, are still in use at modern building sites.

Kinoshita Kōmuten Precut Timber Company, a member of the National

Machine Precut Wooden Residence Association, is typical of a number of small-scale firms which survive in the housing market today through automated joinery machines. It began using these machines in 1978 and is able to produce framing timbers for nearly four houses each day by the use of the new methods. Its output is nearly double that of firms of similar size still using hand tools, and the firm has the additional advantage of much lower training overheads.

The company uses nine different joining and finishing machines on its factory floor. One of these is a multifunction molder imported from West Germany, but the rest have been specially developed in Japan to re-create traditional joinery types without, of course, the traditional joiners. These range from a basic tenon cutter to machines which make the traditional gooseneck mortise for connecting groundsills and beams. The tenoner *(hirahozo-ki)* consists of a series of revolving blades mounted on a steel platform which moves systematically back and forth across the end of the piece of wood. The first blade trims the end neatly before a pair of blades close in on the timber in a pincer movement to fashion the actual tenon. Other machines of similar design fashion dovetailed lap joints *(koshikake-aritsugi)*, large, inset mortise and tenons (*ōiri*), the lapped gooseneck joints *(koshikake-kamatsugi)* used for splicing groundsills, and the complicated rabbeted oblique scarf joint *(okkake-daisentsugi)* for splicing together major beams.

Such manufacturing tools, though extremely ingenious, have some limitations. For instance, the quality of the finish can be quite rough, and experienced carpenters are still employed to finish with chisels certain details which the machines cannot create and to smooth over some of the rougher routing in the large mortises. The precision of the machines also gives rise to some unexpected problems. The "male" and "female" parts of the machine-made joints are identical; thus, should any shrinkage of the wood occur, the interface becomes loose and structurally weak.

The development of joinery machines is a demonstration of the Japanese carpenter doing what he has always done best—creating tools based on long-established principles but adapted to the needs of a particular situation. It is possible that this time he may have created a tool which, while preserving the physical forms of Japanese joinery, will threaten the very existence of the carpenter himself. On the other hand, traditional joinery techniques were already under serious threat as a result of competition from other building systems, and the new joinery tools are certainly stimulating interest in the form and structural performance of traditional joinery. Miyagawa Kōki, a machine tool manufacturer based at Toyohashi in Gifu Prefecture, adapted the shape of the conventional squared tenons and dovetails to create curved edges (fig. 161). This resulted in a significant increase in their structural strength. Their T-shape half-lap dovetail, which is 105 millimeters at the mouth and is used for timber 105 millimeters thick, was certified by the Aichi Prefectural Industrial Standards Council in 1979 with an average breaking strain of 369 kilogram-force. This compares with the 236 kilogram-force for a handmade conventional joint of the same dimensions.

But, one should ask, is new necessarily always better in the case of hand tools and machines in Japan? Even where machines seem to have made some of the most spectacular savings in energy and time, some of the benefits are illusory.

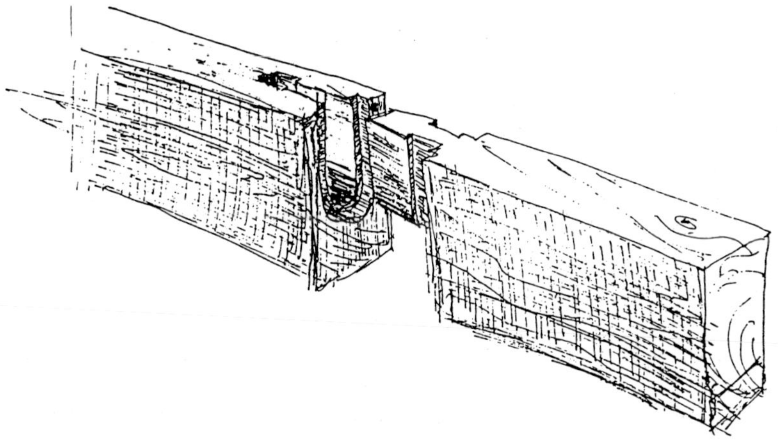

Fig. 161. Machine-made gooseneck mortise-and-tenon (*kama-tsugi*).

The band saw may save much drudgery but, by using the hand ripsaw, the sawyer was able to cut long floor boards and ceiling panels with less wastage than can be done with modern methods. Moreover, traditional methods preserved the natural oil of the timber which is so vital for its durability. Modern power saws, with the high-speed rotation of their cutting blades, actually burn the wood, destroying the oil and sharply reducing its effective life.

At the same time as we recognize the advantages of the techniques and tools used by traditional carpenters, we should not let sentiment cloud reason. Concerns about the advent of the machine and the demise of handcrafts were expressed in the eighteenth century when the spinning jenny was invented in England. In the case of Japan, the mechanization of the building process has assisted the nation to lift its standard of living through higher quality and safer housing. However, many of the developments since 1945 have been short-term responses to necessity. It is now essential to rebuild the infrastructure of the crafts traditions upon which the cultural and economic prosperity of the nation has long rested so they may play a vigorous role in shaping the future. Of all the advanced, industrialized nations, Japan can best afford to take long-term steps for solving the present crisis with creative compromises between hand tools and machines, to discover a new Way of the Carpenter in which tools and trade, man and materials, will again reflect the dynamism and the dignity of the past as they meet the needs of the future.

CHAPTER FIVE

Tools Past and Present, Unique and Universal

This has been the story of Japanese carpenter's tools, the men who made and used them, and the buildings they fashioned. We started with the carpenters and allied building professions, their organization and activity at the building site—in other words, the "who" of the Way of the Carpenter. Later we addressed the "what," "how," "when" and "why" of the tools which link the men, materials and methods in the vocation of building carpentry.

We have also traced the course and probed the logic of Japanese carpentry tools from earliest times to the present. It came as something of a shock to realize, when proper attention is paid to the chronological sequence of tool development, that many of the tools now considered indispensable in traditional building practice came into use relatively late in the development of Japanese architecture. What are now considered to be basic "Japanese tools" were not used to make the historical Japanese buildings whose design and construction techniques have been so extravagantly praised in both Japan and the West. When we look at historic Japanese buildings, such as the Katsura Imperial Villa, we tend to conclude that they were made with tools familiar to us today by assuming that the saws, planes and chisels described in chapter 3 have always been staples of building practice.

Nothing could be further from the truth. Some of the best known Japanese carpentry tools evolved into the form in which they are used today only during the last hundred years. The combination crosscut ripsaw and the plane with capping iron, both considered typical Japanese tool types, are modern inventions. Katsura Imperial Villa and other fine examples of Japanese palatial architecture were created without the assistance of these tools. Other important tools came into existence much later in Japanese history than we might have expected. The block plane was not introduced until the sixteenth century. The use of the ripsaw dates to the fifteenth century, and an effective crosscut saw for rough hewing logs did not come into general use until the seventeenth century.

Thus, erroneous impressions are formed about the way some of the most famous monuments of the Japanese architectural tradition were erected, and false conclusions reached about the meaning of their fundamental design features. As noted in chapter 4, Hōryū-ji was constructed before the advent of many of the standard carpentry tools, making extensive use of hewing axes and adzes, chisels and wedges, planing knives and a preposterously inefficient type of crosscut saw imported from China or Korea. The limited capabilities of the tools available influenced the final appearance of Hōryū-ji as much as any overt stylistic intent of its makers.

Japanese tools came from three sources: importation, invention and a blend of the two. It is this latter process that lies at the heart of much that is both shared and unique in Japanese civilization. It has been shown that many of the basic tools used in Japanese building practice were introduced from the Asian mainland. They were adapted to suit Japanese materials and circumstances, such as the character of the wood and the Japanese preference for sitting down while working. Significant indigenous developments in tool type and use also occurred. The Japanese turned the humble carpenter's square into a versatile instrument of design by adding a scale for working out the curvature of a roof using square roots and tangents. The carpenter's square was also furnished with a scale for determining the dimensions of a squared beam that could be cut from a round log. Metallurgy, associated with the development of the samurai sword in the latter Heian and Kamakura periods, led to remarkable advances in the quality of cutting blades used for chisels and other woodworking tools. Crosscut saws of Chinese origin, which operated on the push stroke, were replaced with saws cutting with a pulling action by the fourteenth century. The two-man frame-saw for ripping lumber, introduced from China before the mid-fifteenth century, was quickly superseded by the single-person maebiki-oga when the demand for castle, palace and religious buildings outstripped the capacity of the two-man team to supply sawn lumber. The plane, when introduced in the sixteenth century from China, was a tool worked by pushing. Less than a century later there had been a rapid proliferation of plane types, from roughing and finishing planes to molding, chamfer and grooving planes. Clumsy handles used for pushing were abandoned, and the more efficient pulling action adopted.

From the mid-nineteenth century, new waves of influence from abroad, this time from Europe and America, once again lead to adoption and adaptation of tools. Plane performance was improved by the addition of the capping iron. A spine was added to the small crosscut saw to make a Japanese version of the tenon saw. At about the same time, standard rip and crosscut saws were combined to create the ryōba as an indigenous development.

In the later part of the twentieth century, the venerated traditions of the Way of the Carpenter have weakened, many of them passing into history. Carpenters now compete with machines for survival in the housing market, but the insistence on the ultimate level of quality still remains the prerogative of the daiku.

Tools have furnished much information about Japanese building practice through the ages, thereby establishing a tangible link with the now intangible processes of building construction. They provide, literally and metaphorically, a handle on the processes of the past and offer deeper understanding of the buildings they shaped and the builders who used them. An understanding of tools also helps correct an overabundant enthusiasm for aesthetics alone in the Western view of traditional Japanese architecture. Although beauty is of profound importance in Japanese architecture, its full appreciation must take place within the broader realm of the Way of the Carpenter—the craftsmen, their tools and their techniques that gave buildings physical reality.

Japanese architecture, therefore, should be seen as craft before art, and tools are the best guide from the craft of building to the art of architecture. Tools are created to enable a craftsman to do a particular job. They act as an extension of his hand and mind, enabling him to transform practical intention and aesthetic desire into physical form. By their very nature they guide the actions of the hand and arm, dictating limits but opening up new possibilities in design and execution. In so doing a tool may be said to shape the mind itself, either overtly or subliminally. This reciprocal relationship between man and his tools is buried deep in the very process of evolution. The anthropologist Sherwood Washburn contends, on the basis of the discovery of protohumanoid bone fragments in Africa, that the invention and use of hand tools stimulated the growth of parts of the brain in prehuman primates.[1] This had profound consequences for physical as well as mental man, ranging from the ability to analyze, articulate and feel emotion, to patterns of child development and physical changes such as smaller teeth because these were no longer used for many of the functions which tools could fulfill. Stone tools such as choppers, scrapers and cutters, however elementary they may seem, significantly expanded man's interior horizons as well as the organization of his exterior world. Tools, in fact, made man as much as man made tools.

Anthropologists may accept such an important connection, but architects and historians have been slower to grasp its implications. Recognition of the ramifications of tool use on the human psyche reveals the limitations of the still current Renaissance-inspired belief that architecture is the product of the mind, and that the process of building is separate from and subservient to the application of theories of formal harmony and proportion. In the Way of the Carpenter, no such dichotomy exists between intellect and application. Design and execution are inseparable; tools are instruments of both the mind and the hand, particularly at the critical creative junctures that arise in the invention of new forms.

Historians generally shy away from analysis of the physical aspects of the building process because of its largely undocumented nature. Trends in architectural theory and practice since the later 1920s and the establishment of the Bauhaus, the seminal school of the modernist movement, have also hindered the appreciation of this issue. The modernist assertion of a universal equation between form and function, prevalent for much of this century, has left little room for pondering the role of process. Subsequently, the rediscovery that buildings serving similar functions are not necessarily identical in form has stirred fresh awareness that process is an active agency in shaping building design. However, misplaced faith in the "mind over matter" adage lingers in our thinking, excluding the possibility that matter may have an influence on the mind, particularly in the case of tools. A major objective of this book has been to identify the role of Japanese carpentry tools in the evolution of Japanese style and the stimulus they gave to the formation of the Japanese artistic mind as expressed in architecture—from curved window mullions to cedar door panels, from wide floor boards to delicate lintels. Necessity may be the mother of invention, but in Japan invention has also fathered necessity. Not only were tools created for specific needs but many tools created for one purpose were found to be useful for other functions as well. This expanded the range of detail and design, in turn enlarging the horizons of human consciousness in that fundamental anthropological sense.

It is important to see Japanese tools not only in their historical setting but also in the universal matrix of tool types and use. There is a tendency to emphasize the uniqueness of Japanese civilization, its isolation and capacity for adaptation of foreign influences to its internal dynamics. The Japanese do this from justifiable pride in their cultural achievements and sometimes from intellectual and linguistic isolation from deep knowledge of other cultures. Many non-Japanese, themselves hindered by lack of language and detailed knowledge of Japanese history and culture, appropriate Japanese architecture into their own personal visions. Western architects such as Frank Lloyd Wright, Bruno Taut and Walter Gropius, while drawing world attention to the significance of traditional Japanese buildings, were seduced into believing that the apparent guileless simplicity of some buildings represented a functionalism akin to that sought in modernist architecture—a mistake akin to that of extolling the unornamented character of Greek classical architecture without the knowledge that time has stripped away the original layers of rich polychrome decoration.

The same process of myth creation is now occurring with Japanese carpenters' tools. Enthusiasm abroad for the superior technical quality of Japanese tools is elevating them to the level of cult objects. Popular woodworking journals invariably carry advertisements for Japanese chisels and saws, while articles record the euphoric reactions of Western tool specialists to their technical capabilities. Such enthusiasm is not misplaced, but the implications of the claim that Japanese tools are unique, which often arise from this enthusiasm, have to be weighed more carefully. It is generally accepted that contemporary Japanese steel blades are without peer among Western tools, but full acknowledgment must also be given to the high standards of steel making before the industrial revolution in Europe and America. Part of the mystique surrounding Japanese saws in the West is because they are pulled not pushed to cut, and planes are drawn toward the body, not pushed away from it. This usage seems less mysterious when the logic of sitting down rather than standing up to perform cutting and smoothing is probed. Much of the mythmaking about uniqueness is simply a result of being uninformed about Japan in general. Zen, sushi, Japanese business management and Japanese education, along with chisels and saws, are all part of a popular view of Japan that mixes self-assured misconceptions with blissful ignorance as part of urban or campus or craft subculture. Within any craft or art, the internal logic of technique, material and purpose is ultimately universal. Drawing together many of the universal attributes of Japanese tools and techniques, which have already been noted in passing, is a useful way of concluding this examination of the Way of the Carpenter.

First, the large number of high-quality tools in traditional Japanese practice and their specialized character parallels the large variety and specialized character of woodworking tools in preindustrial Europe and America, reflected in the old maxim "a tool for every job." In both contexts, this was an inevitable result of trade diversification and the search for greater mechanical efficiency.

Second, the special regard of the Japanese carpenter for his tools, which has a religious dimension, would not come as a surprise to many of the European and American workers of wood of a mere two generations ago. As Walter Rose noted about English circumstances between the wars, "the sawyer's regard for his saw cannot but be something of a mystery. . . . [It] is difficult to understand what it must be like to use a single tool from morning to evening, from end to end of the week. But it is easy enough to believe that such continual use would be reason enough for his devotion to its welfare, his meticulous care to keep it ever in perfect condition." This continual use of a variety of tools, themselves specially crafted to perform a special task for a particular craftsman, employing skills and techniques perfected over many generations and absorbed by the individual carpenter from childhood, helps explain the special devotion of the Way of the Carpenter in Japan in universal terms.

Third, the carpenter's square is a remarkable and versatile tool of design and execution, linking practice and principle. It is very different from the modern set square of the West but is similar in many ways to the European roofer's square with its complicated scales and tables for setting the pitch and position of rafters and purlins. The Japanese and European traditions of timber-frame roofing design had much in common. In both traditions, for example, roof incline was expressed as a number of units rise to units span, not in geometric degrees.

Fourth, long traditions of wedge splitting and axe hewing of structural timbers link both Japan and the west. Japanese, European and American carpenters all recognized the efficiency of splitting straight grained wood with wedges in the making of building timbers and appreciated the longevity, strength and freedom from warping that this process bestowed on the final product.

Fifth, Japanese and European carpenters alike were trained in hereditary practice from an early age. They learned their manual skills as they learned to master the very functioning of their own growing bodies. As Walter Rose again observed, "many are the times, for example, that I have been asked how I know the difference between the various woods—oak, ash, elm, pine, sycamore, beech, and so on. This knowledge I gained so early in life that I do not remember how or when I acquired it." Their thinking about building design matured as their minds themselves matured.

Sixth, methods of working wood in Japan are dictated by universal considerations of wood grain type, direction and density. Fine differences may occur, but the integrity of the relationship between wood character and carpentry technique remains constant. There is a predominance of "softwoods" over "hardwoods"—that is, of conifers over broadleafed, deciduous trees—for structural timber in Japan, but softwoods tend to be the principal timbers used for structural purposes worldwide. There is also a universally shared sense of satisfaction and pride gained in working good wood cleanly and well.

Seventh, arising from the previous point, is the universal love of and sensitivity to wood among the workers of wood. Walter Rose recalls his childhood love for examining newly cut planks and boards propped up for initial drying, "the alcoves thus formed beneath were lofty and cool, even in the hottest weather; the light was subdued within. There was seclusion, complete in its separation from the active life outside. . . . We were sensitive to a subtle influence, as of a Gothic church, to which we likened it; the massed formation of impassive timbers curving upwards, the sweet scent of newly cut wood. We loved to examine those rugged planks and boards and to compare the companion burrs, knots,$_{4}$sloped and grain markings, all taken by man from the beneficent hand of nature, and standing quiet there, in the cool shock, waiting man's further dealings." This deeply spiritual awe of materials and the special role of the woodworker in taking them from nature for man is as much part of European and American traditions as it is part of the Japanese tradition.

Eighth, Japanese and Western saws may now differ in form and use but the bow and frame-saws, once the mainstay of both the European and Asian saw traditions, are similar in form and operational logic. All used a blade kept under tension by a tightened cord. We have already noted how an eyewitness account of British pit-sawyers at work could be mistaken for a description of Muromachi sawyers.

Ninth, the builders of medieval Europe as well as Japan were concurrently designers and builders. Later, as a result of patronage in Europe under the Medici family and the Roman papacy, and in Japan under the three warlords who unified the country—Oda Nobunaga, Toyotomi Hideyoshi and Tokugawa Ieyasu—the architect emerged from the ranks of the artisan class. The types of technical drawings used by architects in each context, although informed by different theories, are remarkably similar in form. Use of modules, as part of systems of design set out in proportion manuals in Japan, is paralleled by the experience of classical and Renaissance practice delineated in the writings and drawings of Vitruvius and Palladio. In both contexts theory was based on well-established practice.

Finally, the circumstances of decline in skills, tool numbers, and loss of prestige and design prerogatives, which are the lot of the Japanese daiku today, unhappily reflect conditions that prevailed in Europe and America between the wars. Mechanization of building construction and the disruption of the patterns of apprenticeship are the reality of the Way of the Carpenter in Japan as in the West.

Carpenters, architects and historians should be encouraged to savor the similarities before delighting in the differences between Japanese and Western tools. Japanese civilization is both unique and universal in character, and it is only by recognizing its dual nature that full understanding becomes possible.

Notes

General acknowledgments and further reading follow the notes for direct quotations in each chapter.

CHAPTER ONE

1. John Ruskin, *Lectures on Art* (London: George Allen, 1889), 61.
2. Iriye Hiroshi, "Apprenticeship training in Tokugawa Japan,"*Acta Asiatica* (Bulletin of the Institute of Eastern Culture) 54 (1988): 1–23.
3. Nishioka Tsunekazu, Takada Kōin, Aoyama Shigeru, *Yomigaeru Yakushi-ji Saito* (The rebirth of the West Pagoda of Yakushi-ji) (Tokyo: Sōshisha, 1981), 212.

THE ETYMOLOGY OF THE TERM DŌGU: *Nippon kokugo daijiten* (Heibonsha, 1975), vol. 14: 440; *Heibonsha daihyakka jiten* (Heibonsha, 1985), vol. 10:599–600.
Before carpenter's tools became known as dōgu, they were referred to as *mokkō* 木工 ("wood working") (Nara period) and *chōdo* 調度 (Heian period), the word which now means utensils and furniture. *Kōgu* 工具, combining the characters for "construction" and "utensil," has been used since the eighteenth century. It appears, for example, in the *Wakan sansaizu-e*, an encyclopedia published in 1712. Although it is still in use today, the term has something of a mechanical flavor and does not convey the richness of meaning of dōgu.

THE APPRENTICESHIP SYSTEM: I am indebted to Endō Motoo, *Nihon shokunin-shi no kenkyū* (Research on the history of Japanese craftsmen) (Tokyo: Yūzankaku, 1961), and *Shokunin no rekishi* (A history of craftsmen) (Tokyo: Shibundō, rev. ed., 1965).

BUILDING CEREMONIES AND RITUALS: Kajima Shuppankai, ed., *Sugu yakunitatsu kenchiku no gishiki to saiten* (Architectural ceremonies and rites which will be immediately useful) (Tokyo: Kajima Shuppankai, 1986).

COURT CUSTOMS AND SHINTO: Felicia G. Bock, trans. and ed., *Engi-shiki: Procedures of the Engi Era, Books I-V* and *Books VI-X* (Tokyo: Sophia University, 1970 [I-V] & 1972 [VI-X]).

SHINTO AND ISE SHRINE: Felicia G. Bock, "The Rites of Renewal at Ise," *Monumenta Nipponica* XXIX-l (Spring 1974): 55–68.

CHAPTER TWO

1. "S'il veut éviter l'ennui des abstractions figées, le livre de l'outil sera également celui des matières et celui des métiers. . . . " André Velter and Marie-José Lamothe, *Le livre de l'outil* (Paris: Denoël/Gonthier, 1976), 1:10.

CHAPTER THREE

1. Republished as Muramatsu Tejirō, ed., *Waga kuni daiku no kōsaku gijutsu ni kansuru kenkyū* ("Research on the building techniques of the *daiku* of Japan") (Tokyo: Rōdō Kagaku Kenkyūjo Shuppanbu, 1984).
2. Muramatsu, *Wagakuni daiku*, 28.
3. Eric Sloane, *A Museum of Early American Tools* (New York: Ballantine Books, 1964), 54.
4. Sloan, *Museum*, 64.
5. Fukuzawa Yukichi, *The Autobiography of Fukuzawa Yukichi*, third rev. ed., trans. Eichi Kiyooka (Tokyo: The Hokuseidō Press, 1947), 10–11.
6. Edward Morse, *Japan Day by Day* (Boston and New York: Houghton Mifflin Company; Cambridge: Riverside Press, 1917), 1:264–5.

CHAPTER FOUR

1. David Ouvry, "The Ancient Pre-fab," *Woodworking* (January 1987), 6–8.
2. "Shoku Nihongi," in *Koji ruien* (Tokyo: Jingū Shichō, 1935–36), 52:1013.
3. Nihon Gakushiin Nihon Kagakushi Kankōkai, ed. *Meijizen Nihon Kenchiku Gijutsu-shi* (A history of Japanese architectural technology before the Meiji period (1961, reprint; Tokyo: Nihon Gakujutsu Shinkōkai, 1982), 203–12.
4. Walter Rose, *The Village Carpenter (1937; reprint, London: A & C Black, 1987)*, 2–4.
5. *Wakan sansaizu-e* (1712; reprint, Tokyo: Yoshikawa Kōbunkan, 1906), 336–40.
6. Regulations and edicts are quoted in Iriye Hiroshi, "Apprenticeship training in Tokugawa Japan," *Acta Asiatica* (Bulletin of the Institute of Eastern Culture), Tōhō Gakkai) 54 (1988), 16.
7. The official records of the Nikko construction project are entitled *Nikkō-san Tōshōdaigongen-sama gozōei onmokuroku* (The official records of the construction on Mount Nikko for Tōshōdaigongen [Tokugawa Ieyasu]) and are kept in the collection of Nikko Tōshō-gū. The interpretation of actual data is difficult for historians because the records, though voluminous, are sometimes unclear.
8. Quoted from an archival manuscript of *Gushi kenki* at the Takenaka Carpentry Tools Museum.
9. Hoshi Akira, *Nihon ongaku kanshō no tebiki* (Handbook for enjoyment of Japanese music) (Tokyo: Ongaku no Tomosha, 1981), 173.
10. Concerning the development of carpentry subprofessions, the first known use of the term *toshōji-shi* occurs in the *Jinrinkinmōzui* (An illustrated compendium of moral admonitions), published in 1690. Tokugawa government edicts first refer to *tateguya-shoku* as a separate category of carpentry in 1793.
11. David John Lu, *Sources of Japanese History*, 2 vols. (New York: McGraw-Hill, 1974), 2: 36.
12. Letter to President Grant on the Iwakura Mission, 1871, in Lu, *Sources*, 2:51. (From the official translation as reproduced in the *New York Times*, March 5, 1872.)
13. Regarding the creation of the *dōtsuki*, the first Western tenon saw was brought to

Japan by Viscount Yamao of Chōshū, who used it while working at shipyards in Glasgow in 1883. This saw is now preserved in the Yokohama Archives of History (Yokohama Kaikōkan).

14. Lu, *Sources*, 2:90. Part of the 1901 Declaration of the Social Democratic Party, drafted by Abe Isoō with help from Katayama Sen and Kōtoku Shūsui, and banned shortly afterwards.

JAPAN PRIOR TO THE ARRIVAL OF BUDDHISM AND ANCIENT ARCHITECTURE: C. Melvin Aikens and Takayasu Higuchi, *Prehistory of Japan*, (New York: Academic Press, 1982); J. Edward Kidder, *The Making of the Past: Ancient Japan* (Oxford: Elsevier-Phaidon Press, 1977).

INFORMATION ON TOOLS FOUND IN *KOFUN*: Takenaka Daiku Dōgukan, ed., *Takenaka Daiku Dōgukan tenji kaisetsu* (Explanation of the exhibits of the Takenaka Carpentry Tools Museum) (Tokyo: Nihon Shashin Insatsu, 1984); Karai Kunio, "Nihon no daiku dōgu—nomi" (Japanese carpenter tools—chisels) *Zenken jānaru* (All-building Journal) 10 (1986), 44–47.

ISE AND IZUMO SHRINES: Yasutada Watanabe, *Ise and Izumo*, trans. by Robert Ricketts (New York and Tokyo: John Weatherhill, 1974); Inagaki Eizō, "Shikinen sengū no kenchikuteki kōsatsu"(Architectural observations on the periodic rebuilding of shrines) in *Nihon kenchiku no tokushitsu—Ōta Hirotarō hakase kanreki kinen rombunshū* (Special character of Japanese architecture—essays commemorating Dr. Ōta Hirotarō's sixtieth birthday) (Tokyo: Chūōkōron Bijutsu Shuppan, 1976), 189–212.

THE BRONZE RAISED FLOOR STRUCTURE: *Wen wu* (Chinese Journal of Archaeology) 5 (1964), 33–40.

HŌRYŪ-JI CONSTRUCTION TOOLS: Karai Kunio, "Daiku dōgu no hanashi, Takenaka Daiku Dōgukan shūshūhin yori—daikuyō noko"(Stories about Japanese carpenters' tools from the collection of the Takenaka Carpentry Tools Museum—the carpenter's saw) *Shahō Takenaka* (Takenaka Company Bulletin) (1983); Karai Kunio, "Hōryū-ji kennō hōmotsu no noko" (The saw in the Hōryū-ji Treasury), "Nihon no daiku dōgu 1," *Zenken jānaru* (January 1986), 46–49; Takeshima Takuichi, "Hōryū-ji no kōgu" (Woodworking tools of Hōryū-ji), *Sekai bijutsu zenshū* (Complete collection of world art), vol. 2 (Tokyo: Kadokawa Shoten, 1963), 37–46; Hōryū-ji shūri kōji iinkai, ed., *Kokuhō Hōryū-ji Kondō shūri kōji hōkokusho* (Restoration report for the Main Hall of Hōryū-ji) (Nara: Shinyōsha, 1956); Nihon Gakushiin Nihon Kagakushi Kankōkai, ed., *Meijizen Nihon kenchiku gijutsu-shi*, 228–29; Miyakami Shigetaka and Hozumi Kazuo, *Hōryū-ji, sekai saiko no mokuzō kenchiku* (Hōryū-ji, the world's oldest timber-frame architecture), vol. 1 of *Nihonjin wa dono yō ni kenzōbutsu o tsukutte kita ka* (How did the Japanese make buildings?) (Tokyo: Sōshisha, 1980).

THE NARA PERIOD: Nara rokudai-ji taikan henshūbu, ed., *Nara rokudai-ji taikan* (General collection of the six great temples of Nara), vol. 9–11 [Tōdai-ji, 1–3] (Tokyo: Iwanami Shoten, 1968–19); Ōta Hirotarō, "Kodai kenchiku no seisan" (Production of ancient architecture) in *Nara rokudai-ji taikan*, Tōdai-ji 1, appendix VI, 1–12; Itō Nobuo et al., *Nihon koji bijutsu zenshū : Tōdai-ji to Shinyakushi-ji, Hokke-ji*, (Complete collection of Japanese ancient art: Tōdai-ji and Shinyakushi-ji, Hokke-ji) vol. 4 (Tokyo: Shūeisha, 1980); Asano Kiyoshi, *Nara jidai kenchiku no kenkyū* (Research on architecture of the Nara period) (Tokyo: Chūōkōron Bijutsu Shuppan, 1969); Miyamoto Nagajirō and

Hozumi Kazuo, Heijō-kyō, *Kodai toshi, keikaku to kenchiku* (Heijō capital, ancient city planning and architecture) vol. 7 of *Nihonjin wa dono yō ni kenzōbutsu o tsukutte kita ka* (1986); William H. Coaldrake, "The Architecture of Tōdai-ji," in John M. Rosenfield et al., *The Great Eastern Temple: Treasures of Japanese Buddhist Art from Tōdai-ji* (Bloomington: Indiana University Press, 1986), 33–47.

NARA-PERIOD MEASUREMENTS: Koizumi Kesakatsu, *Monosashi: Mono to ningen no bunka-shi* (Measuring: cultural history of people and objects series), vol. 22 (Tokyo: Hōsei Daigaku Shuppankyoku, 1977), 60–78; Mochizuki Nagayo, *Nihonjin no shakudo* (Shaku measuring system of the Japanese) (Tokyo: Rokugei Shobō, 1971), 161–254.

EMAKIMONO AND CONSTRUCTION SCENES: Tanaka Ichimatsu, ed., *Nihon emakimono zenshū* (Complete collection of Japanese narrative picture scrolls), 24 vols. [esp. vol. 15, *Kasuga Gongen Genki-e*] (Tokyo: Kadokawa Shoten, 1958–69); Itō et al., *Ishiyama-dera to Ōmi no koji : Nihon koji bijutsu zenshū* (Ishiyama-dera and old temples of Ōmi Province: complete collection of Japanese ancient art) vol. 11 (1981); Okudaira Hideo, *Narrative Picture Scrolls: Arts of Japan*, vol. 5, trans. and adapt. Elizabeth ten Grotenhuis (New York and Tokyo: John Weatherhill, 1973).

THE HISTORY OF JOINERY: Iwate Tamotsu, "Kama-tsugi ni tsuite: keishiki no hensen to jissoku sūchi ni yoru tōkeigakuteki kenkyū" (Changes in the form of the gooseneck mortise-and-tenon joint, based on statistical analysis of actual examples), *Kenchikushi kenkyū* (Research in architectural history) 36 (4/1965), 16–34.

ROOF STRUCTURE AND EVOLUTION: Mary Neighbour Parent, *The Roof in Japanese Buddhist Architecture* (New York and Tokyo: John Weatherhill, 1983).

ANALYSIS OF SONG ARCHITECTURE: Liang Ssu-ch'eng, *A Pictorial History of Chinese Architecture*, ed. Wilma Fairbank (Cambridge: MIT Press, 1984).

MEDIEVAL SAWS: Nihon gakushiin Nihon kagakushi kankōkai, ed., *Meijizen Nihon kenchiku gijutsu-shi*, 225–39; Karai Kunio, "Chūsei no daikuyō noko (konohagata-noko), Nihon no daiku dōgu 2" (Medieval carpenters' saws [the leaf-blade saw]), Japanese carpentry tools 2), *Zenken jānaru* (February 1986), 40–43; Hirasawa Kazuo, *Nokogiri* (The saw) (Tokyo: Kuori, 1980), 106–9 [early written references to *oga*],120–22 [information about Takane Shrine and the Hōjō order to the master sawyers of Fujisawa]; Muramatsu Teijirō, *Daiku dōgu no rekishi* (A history of carpentry tools) (Tokyo: Iwanami Shinsho, 1973), 46–55 [Muramatsu notes that the Chinese use the splayed arrangement of teeth for their frame-saws but that frame-saws still used today in isolated regions of the Korean peninsula have teeth pointing in the same direction, see p. 50]; Yoshikawa Kinji, *Nokogiri—Mono to ningen to bunka-shi* (Saws—a cultural history of people and objects series) 18 (Tokyo: Hōsei Daigaku Shuppankyoku, 1976), 79–149; Karai Kunio, "Nihon no daiku dōgu" (Japanese carpentry tools) *Zenken jānaru* (May 1986), 42–45, Ibid., "Chūsei no seizaiyō nokogiri 1–2" (Medieval lumber saws, 1–2) (June 1986), 44–47, Ibid., "Kinsei ikō no seizai nokogiri" (Lumber saws after the medieval period) (July 1986), 44–47, Ibid., "Kinsei ikō no batsuboku zōzai nokogiri" (Felling saws of the Momoyama and Edo periods) (August 1986), 44–47.

THE SHAKUBUJI *OGA*: Hirasawa, *Nokogiri*, 106–109; Karai, "Nihon no daiku dōgu," *Zenken jānaru* (May 1986), 44–45.

THE KUSADOSENGENCHŌ LEAF-BLADE CROSSCUT SAW: Matsushita Masashi, "Kusadosengenchō iseki" (Kusadosengenchō site), *Nihon no bijutsu* 215 (April 1984).

COMPARATIVE INFORMATION FROM EUROPE: Charles Singer et al., eds., *A History of Technology*, 5 vols. (Oxford: Clarendon Press, 1958), 2:106–23, 388–95.

THE RYŌGEN-IN OF DAITOKU-JI: Itō et al., *Nihon koji bijutsu zenshū*, vol. 13 [Daitoku-ji]; Bruce A. Coats, "The Architecture of Zen-sect Buddhist Monasteries in Japan 1200–1500," Ph.D. diss., Harvard University, 1985, vol.1, 329–35.

THE ANRAKU-JI THREE-STORY PAGODA: Nagano-ken kyōiku iinkai, ed., "Anraku-ji," in Nagano-ken bunkazai hogo kyōkai, ed., *Nagano-ken no kokuhō jūyō bunkazai kenzōbutsu shūri kōji hōkokusho* (Restoration reports for national treasures and important cultural properties in Nagano Prefecture) 1 (1958, reprint; Matsumoto: Kyōdo shuppansha, 1988), 353–400.

THE MOMOYAMA AND EDO PERIODS: Nishi Kazuo, *Edo jidai no daikutachi* (Carpenters of the Edo period) (Kyoto: Gakugei Shuppansha, 1982); Endō Motoo, *Shokunin no rekishi* (History of craftsmen) rev. ed. (Tokyo: Shibundō, 1965), 48–75 [*shoku-nakama*]; William H. Coaldrake, "Gateways of Power: Edo Architecture and Tokugawa Authority 1603–1651," Ph.D diss., Harvard University, 1983; William H. Coaldrake, "Edo Architecture and Tokugawa Law," *Monumenta Nipponica*, XXXVI–3 (1981), 235–84; Ōta Hirotarō and Itō Yōtarō, eds., *Shōmei* (Elucidation of craft) 2 vols. (Tokyo: Kajima Shuppankai,1971); Tanabe Yasushi, "Edo bakufu daitōryō shokusei ni tsuite" (The employment system for chief master carpenters in the Tokugawa shogunate), *Kenchiku zasshi* (Architecture journal) 598 (1935), 27–33; Tanabe Yasushi, "Edo bakufu daitōryō Kōra-shi ni tsuite" (Chief Master Carpenter Kōra of the Tokugawa shogunate), *Kenchiku zasshi* 609 (1936), 123–29; Naitō Akira et al., eds., *Rakuchū rakugai-zu—Funaki-ke kyūzōbon* (Scenes in and around Kyoto—the screens formerly in the collection of the Funaki family) (Tokyo: Shōgakukan, 1987); The Seibu Museum of Art and The Nihon Keizei Shimbun, eds., *Kawahara Keiga-ten* (Catalogue of the exhibition of Kawahara Keiga) (Tokyo: Shōgakukan, 1987).

THE MEIJI PERIOD AND MODERN ERAS: Horikoshi Saburō, *Meiji shoki no yōfū kenchiku* (Western-style architecture in the early Meiji period) (1929; reprint, Tokyo: Nanyōdō, 1973); Ardath W. Burks, *The Modernizers—Overseas Students, Foreign Employees and Meiji Japan* (London: Westview Replica Editions, 1985); Basil Hall Chamberlain, *Things Japanese, being notes on various subjects connected with Japan for the use of travellers and others*, 5th ed. rev. (London: John Murray, 1905); R. T. Chang, *From Prejudice to Tolerance: 1826–1864* (Tokyo: Sophia University Press, 1970); Fujii Jintarō, comp. and ed., *Outline of Japanese History in the Meiji Era*, trans. and adapt. H. K. Colton and K. E. Colton (Tokyo: Ōbunsha, 1958); Carol Gluck, *Japan's Modern Myths: Ideology in the late Meiji Period* (Princeton: Princeton University Press, 1985); Hasegawa Takashi et al., *Nihon no kenchiku—Meiji, Taishō, Shōwa* (Japanese architecture—Meiji, Taishō and Shōwa eras), 10 vols. (Tokyo: Dekoboko Insatsu, 1983); Marius B. Jansen and Gilbert Rozman, eds., *Japan in Transition: From Tokugawa to Meiji* (Princeton: Princeton University Press, 1986); Muramatsu Teijirō, *Nihon kindai kenchiku no rekishi* (Modern Japanese architecture) (Tokyo: NHK Books, 1977); Uyeno Naoteru, ed., *Japanese Arts and Crafts in the Meiji Era*, adapt. Richard Glain (Tokyo: Pan-Pacific Press, 1958); Taniguchi Yoshirō, *Hakubutsukan Meiji-mura* (Meiji Village Museum) (Tokyo: Tankōsha, 1976).

TOKYO URBAN PLANNING: Fujimori Terunobu, *Meiji no Tōkyō no keikaku* (The planning of Meiji-period Tokyo) (Tokyo: Iwanami Shoten, 1982); William H. Coaldrake, "Order and Anarchy: Tokyo from 1868 to the Present," in Mildred Friedman, ed., *Tokyo: Form and Spirit* (New York: Harry N. Abrams, 1986), 62–75.

EDUCATIONAL REFORM: Ishikawa Matsutarō, "The Meiji Restoration and Educational Reforms," *Acta Asiatica* (Bulletin of the Institute of Eastern Culture, Tōhō Gakkai) 54 (1988), 24–47.

THE SŌGAKUDŌ: Zaidan hōjin bunkazai kenzōbutsu hozon gijutsu kyōkai, ed., *Kyū Tōkyō Ongakkō Sōgakudō ichiku shūri kōji hōkokusho* (Record of the resiting and restoration of the Sōgakudō of the former Tokyo School of Music) (Tokyo: Shinyōsha, 1987); Tokyo Shimbun Shuppankyoku, ed., *Ueno Sōgakudō monogatari* (Tale of the Ueno Sōgakudō) (Tokyo: Tokyo Shimbun Shuppankyoku, 1987).

THE CULTURAL PRESERVATION LAWS: Itō Nobuo, "Bunkazai kenzōbutsu hogo jigyō no gaiyō" (Outline of architectural properties preservation work) *Kenchiku zasshi*, (December 1959), 1–10; Bunka-chō Archives contain copies of Cultural Properties Preservation Laws (Koshaji hozonhō, 1903; Koshaji hozonhō shikō ni kansuru ken, 1903).

THE RESTORATION OF TŌDAI-JI: Tōdai-ji Daibutsuden Shōwa daishūri shūri iinkai, ed., *Kokuhō Tōdai-ji Kondō (Daibutsuden) shūri hōkokusho* (Record of restoration of the Great Buddha Hall of the National Treasure Tōdai-ji) 3 vols. (Nara: Meishinsha, 1980); Nara Rokudaiji Taikan Kyōkai, ed., *Nara rokudaiji taikan* vol. 9 [Tōdai-ji 1] (Tokyo: Iwanami Shoten, 1968–72); William H. Coaldrake, "Architecture at Tōdai-ji," in John M. Rosenfield et al., *The Great East Temple: Treasures of Japanese Buddhist Art from Tōdai-ji* (Bloomington: Indiana University Press, 1986), 32–47.

REFORMS TO WEIGHTS AND MEASURES: Koizumi Kesakatsu, *Monosashi, Mono to ningen no bunkashi* (Measures, the cultural history of people and objects) vol. 22 (Tokyo: Hōsei Daigaku Shuppankyoku, 1977).

THE MOVEMENT TO PRESERVE AND REVITALIZE THE TRADITIONAL BUILDING PROFESSIONS: William H. Coaldrake, "The Resurrection of the *Shokunin*—The Third Conference of the Timber-frame Architecture Research Forum, Kyoto, February, 1987," *The Japan Architect* (May 1987), 6–7.

INDUSTRIALIZED HOUSING: William H. Coaldrake, "Manufactured Housing—The New Japanese Vernacular," *The Japan Architect*, four-part series (August-November 1986, and January 1987). The authoritative source for statistical data on the contemporary housing industry is the annual White Paper on the Housing Industry (*Jūtaku sangyō hakusho*) issued by Yano Keizai Kenkyūsho (2-10-1 Hama-cho, Nihonbashi, Chuo-ku, Tokyo). Close to one thousand pages in length, it summarizes information from housing industry sources for the preceding fiscal year and analyzes long-term trends.

CHISEL QUALITY: Bill Stankus, "Testing Wood Chisels: Lab Finds No Secrets in the Steel," *Fine Woodworking* no. 51 (March/April 1985), 44–48.

CHAPTER FIVE

1. Sherwood L. Washburn, "Tools and Human Evolution," *Scientific American* (September 1960). Reprinted in Glynn Isaac and Richard E. F. Leakey, eds., *Human Ancestors* (San Francisco: W. H. Freeman & Company, 1979), 9–21.
2. Walter Rose, *The Village Carpenter*, 6.
3. Ibid., 1.
4. Ibid., 3–4.

Guide to Further Reading

General Works in English on Japanese Architecture and Building Practice

Coaldrake, William, H. "Myths and Realities of Japanese Building" *Architecture* (September 1988): 113–17.

Hashimoto, Fumio. *Architecture in the Shoin Style, Japanese Feudal Residences.* Translated and adapted by H. Mack Horton. Tokyo: Kodansha International, 1981.

Hinago, Motoo. *Japanese Castles.* Translated and adapted by William H. Coaldrake. Tokyo: Kodansha International, 1986.

Morse, Edward S. *Japanese Homes and their Surroundings.* 1885. Reprint. Rutland, Vermont and Tokyo: Charles E. Tuttle, 1972.

Nakamura, Yasuo. *Japanese Joinery: Handbook for Joiners and Carpenters.* Translated by Koichii Paul Nii. Point Roberts, Washington: Cloudburst Press Book/Hartley and Marks, 1983.

Nishi, Kazuo and Kazuo Hozumi. *What is Japanese Architecture? A survey of traditional Japanese architecture, with a list of sites and a map.* Translated, adapted, and with an introduction by H. Mack Horton. Tokyo: Kodansha International, 1985.

Ōta, Hirotarō, ed. *Traditional Japanese Architecture and Gardens.* Translated by Kirishiki Shinjiro. Yokohama: Kokusai Bunka Shinkōkai, 1972.

Seike, Kiyosi. *The Art of Japanese Joinery.* Translated and adapted with an introduction by Yuriko Yobuko and Rebecca M. Davis. New York and Tokyo: John Weatherhill, 1977.

Suzuki, Kakichi. *Early Buddhist Temple Architecture in Japan.* Translated and adapted by Mary Neighbour Parent and Nancy Shatzman. Tokyo: Kodansha International, 1980.

Specialized Studies in Japanese on the Building Professions

Endō, Motoo. *Nihon shokuninshi no kenkyū* (Research on the history of Japanese craftsmen). Tokyo: Yuzankaku, 1961. Studies on the building craftsmen up to the sixteenth century.

_________ . *Shokunin no rekishi* (A history of craftsmen). 1956. Revised. Tokyo: Shibundo, 1965. Concentration on the Momoyama and Edo periods.

Shinkenchiku-sha, ed. *Ezu daiku hyakutai* (A hundred pictures of carpenters at work) 2 vols. in English-Japanese. Tokyo: Shinkenchiku-sha, 1974.

Specialized Studies in Japanese on Carpentry Tools

Muramatsu Teijirō. *Daiku dōgu no rekishi.* (The history of Japanese carpenters' tools) Tokyo: Iwanami Shinsho, 1973. The first major study of the history of Japanese carpenters' tools. Some of the details and conclusions are now out of date.

Muramatsu Teijirō, ed. *Waga kuni daiku no kōsaku gijutsu ni kansuru kenkyū* (Research on the building techniques of Japanese carpenters) 1949. Reprint. Tokyo: Rōdō Kagaku Kenkyūjo Shuppanbu, 1984. The 1943–44 Institute of Labor and Science survey of Japanese carpentry tools.

Nihon Gakushiin Nihon Kagakushi Shinkōkai, ed. *Meijizen Nihon kenchiku gijutsushi.* (A history of Japanese architectural technology before the Meiji period). 1961. Reprint. Tokyo: Nihon Gakujutsu Shinkokai, 1982. See especially pp.187–251.

Shinkenchiku-sha, ed. *Daiku dōgushū* (A collection of Japanese carpenters' tools).Tokyo: Shinkenchikusha, 1984. The best illustrated description of the main types of tools.

Introductory Studies in English on Carpentry Tools

Lanz, Henry. *Japanese Woodworking Tools—Selection Care and Use*. New York: Sterling Publishing Company, 1985. The book emphasizes tool making and preparation for use.

Ōdate, Toshio. *Japanese Woodworking Tools: Their Tradition, Spirit and Use.* Newtown, Connecticut: The Taunton Press, 1984. An introduction to the woodworking tools and techniques of the *tateguya,* the screen-maker.

Special Collections of Japanese Carpentry Tools

The Takenaka Carpentry Tools Museum (Takenaka Daiku Dōgukan) in Kobe, Japan, has extensive displays and holdings of tools, and the staff of the museum has prepared a comprehensive set of videotapes showing how each tool is used and made. In the United States, there is a fascinating collection of carpenters' tools, dating from the later nineteenth century in the Morse Collection at the Peabody Museum in Salem, Massachusetts.

Studies of Western Tools

Bertorelli, Paul, ed. *Fine Woodworking on Hand Tools: 38 Articles Selected by the Editors of Fine Woodworking Magazine*. Newtown, Connecticut: The Taunton Press, 1986.

Singer, Charles et al., eds. *A History of Technology*. 5 vols. Oxford: Clarendon Press, 1958.

Sloane, Eric. *A Museum of Early American Tools*. 1956. Reprint. New York: Ballantine Books, 1973.

Velter, André and Marie-José Lamothe. *Le livre de l'outil*. 2 vols. Paris: Denoël/ Gonthier, 1976.

Glossary

TOOLS

anazuki-ganna (穴突鉋), grooving plane
arashiko-ganna (荒仕工鉋), general roughing plane
awase-jōgi (合せ定規), plane sole gauge
azehiki-noko (畔挽鋸), see *hanamaru-noko*
chōna (釿，手斧), adze
chūshiko-ganna (中仕工鉋), intermediate smoothing plane
dai-ganna (台鉋), plane, "block plane"
daigiri-oga (台切大鋸), crosscut lumber saw
dōtsuki-noko (胴付鋸), tenon or back saw
ezu-ita (絵図板), plan board
gagari (がゞり，鋸賀利), heavy-duty carpenter's ripsaw
gennō (玄翁), steel double-faced hammer
hakogata-jōgi (箱型定規), block gauge
hanamaru-noko (鼻丸鋸), "rounded nose" roughing saw
hidensho (秘伝書), "secret hereditary writings"; carpenter's design manual
hikimawashi-noko (挽廻鋸), coping saw
hinagata-bon (雛形本), "pattern book;" carpenter's design manual
hira-ganna (平鉋), smoothing plane
hiro-ṇomi (広鑿), striking chisel with head wider than 3 centimeters (1 *sun*)
hoso-noko (細鋸), narrow saw used for cutting slots for plane blade
ita-zu (板図), see *ezu-ita*
jōgi (定規), measuring gauge
jōshiage-ganna (上仕上鉋), special finishing plane
kakumentori-ganna (角面取鉋), beveling or chamfer plane
kama-kebiki (鎌毛引), double-line marking gauge. See also *nimai-kebiki.*
kanazuchi (金槌), carpenter's hammer
kanejaku (曲尺), see *sashigane*
kanna (鉋), plane ; see also *dai-ganna*
kebiki (毛引), marking gauges
kensao (間竿), long measuring rod
kiri (錐), gimlet

kiwa-ganna (際鉋), corner-cutting plane
kizuchi (木槌), wooden mallet
konohagata-noko (木の葉形鋸), "leaf-blade" crosscut saw developed by the twelfth century. Evolved into the pull-action crosscut saw in the fifteenth century.
kote-nomi (鏝鑿), "trowel" chisel
kugi-nuki (釘抜), nail extractor
maebiki-noko (前挽鋸), standard purpose carpentry ripsaw
maebiki-oga (前挽大鋸), wide-bladed ripsaw
masakari (鉞), hewing axe
mentori-ganna (面取鉋), see *kakumentori-ganna*
motoichi-jakuri-ganna (基市決鉋), fenced plough plane
nagadai-ganna (長台鉋), jack plane
nimai-kebiki (二枚毛引), double-line marking gauge
nokogiri (鋸), saw
nomi (鑿), chisel
oga (大鋸), literally, "big saw;" the frame-saw, usually operated by two men, used for ripsawing lumber. Introduced from China in the fifteenth century.
ōire-nomi (追入鑿), butt chisel
oni-arashiko-ganna (鬼荒仕工鉋), preliminary roughing plane
osa-jōgi (筬定規), pillar profile gauge; also known as *osa maruta-biki*
ryōba (両歯), double-edged saw
sagefuri (下げ振り), plumb line
sashigane (曲尺), carpenter's square
shakuzue (尺杖), see *kensao*
shiage-ganna (仕上鉋), special finishing plane
sokosarai-nomi (底渫鑿), "bottom cleaning" chisel
sokotori-jakuri-ganna (底取决鉋), "bottom shaving" plough plane
soridai-ganna (反り台鉋), molding plane
sotomaru-ganna (外丸鉋), concave groove molding plane
suji-kebiki (筋毛引), marking gauge
sumisashi (墨芯), bamboo inking pen
sumitsubo (墨壺), ink pot and snap line
tataki-nomi (叩鑿), striking chisel
tatebiki-noko (縦挽鋸), ripsaw
toishi (砥石), whetstone
tomegata-jōgi (留型定規), miter joint gauge
tsuba-nomi (鐔鑿), "sword-guard" chisel
tsuki-nomi (突鑿), paring chisel, slick
uchidashi-nomi (打出し鑿), "strike-through" chisel
uchikomi-giri (打込み錐), striking gimlet
uchimaru-ganna (内丸鉋), convex groove molding plane
waki-ganna (脇鉋), side-shaving plane
wari-kebiki (割毛引), splitting gauge
yari-ganna (槍鉋), planing knife
yasuri (鑢), files, especially for sharpening sawteeth
yokobiki-noko (横挽鋸), standard crosscut saw

GENERAL ARCHITECTURAL AND JAPANESE TERMINOLOGY

akamatsu (赤松), *Pinus densiforia*, red pine

bakufu (幕府), literally, "tent government." Shogunate, first by the Minamoto at Kamakura (1185–1333), then by the Ashikaga at Muromachi in Kyoto (1333–1573), and finally by the Tokugawa at Edo (1603–1868).

banshō (番匠), term for carpenter used from Nara period (710–794) into medieval times

banzuke (番付け), a system of labeling and marking timbers

bu (分), traditional unit of measure equivalent to 10 *rin*, or approximately 3.03 millimeters

chintaku reifu (鎮宅霊符), Shinto god who protects carpenters and their work

daiku (大工), general term for carpenter; master carpenter. Originally title for highest rank of artisan, meaning "great craftsman."

hinoki (檜), *Chamaecyparis obtusa*, Japanese cypress

ibara-me (茨目), "thorn" shaped tooth; the raked tooth used for ripsaws

jichinsai (地鎮祭), ceremony for the gods of the earth, ground-breaking ceremony

jō (丈), traditional unit of measure equivalent to 10 *shaku*, or approximately 3.03 meters

jōtōshiki (上棟式), ridge-raising ceremony, marking the completion of frame assembly. Also known as *mune-age-shiki.*

kama-tsugi (鎌継), gooseneck mortise-and-tenon joint

kara-hafu (唐破風), cusped gable. Most elegant form of ornamental roof

ken (間), traditional unit of measure equivalent to 6 *shaku*, or approximately 1.82 meters

keyaki (欅), *Zelkova accuminita*, zelkova

kiwari (木割), literally, "timber splitting." A system of proportion design based on modules, such as the span of the bay.

kobiki (木挽き), sawyer. See also *oga-hiki.*

kuromatsu (黒松), *Pinus thunbergi*, black pine, common pine

miya-daiku (宮大工), carpenter specializing in temple and shrine buildings

munafuda (棟札), ridge plaque bearing prayers for the safety of the building and listing principal participants in its construction

muneageshiki (棟上式), see *jōtō-shiki*

nageshi (長押), horizontal beam which wraps around the pillar. Originally braced the structure, but later became purely decorative.

nuki (貫), penetrating tie-beam used to secure the building frame horizontally

oga-hiki (大鋸挽き), early name for sawyer

rakuseishiki (落成式), ceremony marking the completion of the building project

rin (厘), traditional unit of measure equivalent to approximately 0.303 millimeters or one-tenth of a *bu*

sashimonoshi (指し物師), carpenter specializing in wooden containers made with intricate joints; "joiner"

shaku (尺), traditional unit of measure equivalent to 10 *sun*, or approximately 30.3 centimeters

shiguchi (仕口), angled or right-angled joint

shikii (敷居), sill

shinden-zukuri (寝殿造), style of mansion architecture developed during the Heian period (794–1185) for members of the aristocratic class in the city of Heian (Kyoto)

shoin-zukuri (書院造), style of residential architecture which reached maturity at the end of the sixteenth century. Characterized by asymmetrical grouping of buildings in landscaped setting, with interior chambers equipped with tatami (畳) mats, tokonoma (床の間) and split-level shelving (*chigaidana* 違い棚) as well as gold-backed screens and wall paintings.

sugi (杉), *Cryptomeria japonica*, Japanese cedar

sukiya-daiku (数寄屋大工), carpenter specializing in *sukiya* ("teahouse-style") buildings

sukiya-zukuri (数寄屋造), "teahouse-style" architecture that evolved in the sixteenth and seventeenth centuries; buildings combining *shoin-zukuri* elements with the aesthetics of refinement and conscious rusticity associated with the tea-ceremony tradition

sun (寸), traditional unit of measure equivalent to 10 *bu*, or approximately 3.03 centimeters

tateguya (建具屋), carpenter specializing in screen frames and doors

tōryō (棟梁), literally, "ridgepole;" chief master carpenter in charge of building site

tsubo (坪), traditional unit of measure equivalent to the area of two tatami mats laid side by side, or approximately 3.305 square meters

tsugite (継手), spliced or scarf joint

urame (裏目), square root scale on rear of long arm of carpenter's square. Used for calculating curvature and for determining sectional dimensions of a squared beam obtainable from an unhewn log.

Index